Don't Worry
IT'S SAFE TO EAT

To Arran, Elwen and Freya
I hope you still have a choice in what you eat when you are older

Don't Worry
IT'S SAFE TO EAT

THE TRUE STORY OF GM FOOD, BSE & FOOT AND MOUTH

ANDREW ROWELL

Earthscan Publications Limited

London • Sterling, VA

First published in the UK and USA in 2003 by
Earthscan Publications Ltd

ISBN: 1-85383-932-9

Typesetting by MapSet Ltd, Gateshead, UK
Printed and bound by Creative Print and Design (Wales), Ebbw Vale
Cover design by Ruth Bateson

For a full list of publications please contact:

Earthscan Publications Ltd
120 Pentonville Road
London, N1 9JN, UK
Tel: +44 (0)20 7278 0433
Fax: +44 (0)20 7278 1142
Email: earthinfo@earthscan.co.uk
Web: **www.earthscan.co.uk**

22883 Quicksilver Drive, Sterling, VA 20166–2012, USA

A catalogue record for this book is available from the British Library

Library of Congress Cataloging-in-Publication Data

Rowell, Andrew.
    Don't worry, it's safe to eat : the true story of GM food, BSE, and Foot and
    Mouth / Andrew Rowell.
        p. cm.
    Includes bibliographical references and index.
    ISBN 1-85383-932-9 (hardback)
      1. Genetically modified foods–Great Britain. 2. Bovine spongiform
    encephalopathy–Great Britain. 3. Foot-and-mouth disease–Great Britain.
    I. Title.

    TP248.65.F66R69 2003
    363.19'2'0941–dc21

                                                        2003006473

Earthscan is an editorially independent subsidiary of Kogan Page Ltd and
publishes in association with WWF-UK and the International Institute for
Environment and Development

This book is printed on elemental chlorine-free paper

Contents

Preface

In March 2003, the Speaker of the US House of Representatives, Dennis Hastert, testified before an agriculture committee in Washington. 'Over the last few years, we have seen country after country implementing protectionist, discriminatory trade policies under the cloak of food safety' said Hastert, 'each one brought on by emotion, culture, or their own poor history with food safety regulation.'

Hastert went on to call on the USA to take the European Union (EU) to the World Trade Organization (WTO) due to the latter's moratorium on genetically modified (GM) food. Hastert, like many in the Bush administration, see this as an illegal barrier to trade, against a safe, proven technology.

Hastert continued that farmers had been modifying plants since the 'dawn of time', adding that 'biotechnology is merely the next stage of development in this age-old process'. He contended that 'the European Union has had an indefensible moratorium on genetically modified products in place for over four years with no end in sight. This is a non-tariff barrier based simply on prejudice and misinformation, not sound science. In fact, their own scientists agree that genetically modified foods are safe'.

Despite what Hastert would like us to believe, biotechnology is not simply an extension of plant breeding. Moreover, many scientists do not believe that GM products have been proven to be safe to the environment and humans. People's concerns about GM are not based on misinformation. But it is misinformation from government and scientists over issues such as BSE that means that people do not trust politicians and scientists anymore when they tell them that something is safe to eat.

Hastert's comments reflect those of a Bush administration that is determined to force people to eat GM food, whereas we know that millions of people do not want to eat GM and demand a choice in what they eat.

In the UK there is a 'public debate' on GM, although many people believe it could be a charade. All the signs are that the Labour government want GM commercialization and the biotech industry regards commercialization as inevitable. Even if the government listened to consumers, they could be powerless to stop GM due to probable action by the USA at the WTO.

If people in Europe are to resist the enormous economic and political pressure to commercialize GM then they need to be informed. This book is intended to help people make up their own minds about GM, as well as the lingering issue of BSE.

Some will argue that this book is scaremongering. It is certainly not. It is not intended to be an 'anti-GM' book – more one that raises issues and concerns that must be addressed if we are to protect public health. I argue it is time to implement safe food policies that will avert another food disaster, and this book offers radical solutions for farming and science that would provide safe food and common sense science.

As with much of the science covered in this book, there is great disagreement. Some of the scientists that are interviewed in the following pages disagree with each other. So just because they are interviewed, it does not mean they support the views of other scientists. But they all agree that we need an honest, open and inclusive debate about food safety and the future of science. Hopefully this book will help spark that debate, before it is too late.

STOP PRESS

Just as this book was going to press in May 2003, the US Trade Representative Robert Zoellick and Agriculture Secretary Ann Veneman announced that the USA, Argentina, Canada and Egypt would file a case at the WTO against the EU's 'illegal' five-year moratorium on genetically modified organisms (GMOs).

The USA was joined by Australia, Chile, Colombia, El Salvador, Honduras, Mexico, New Zealand, Peru and Uruguay, but not by African nations that it had been trying desperately to get on board. 'The EU's moratorium violates WTO rules', said Zoellick. 'People around the world have been eating biotech food for years. Biotech food helps nourish the world's hungry population, offers tremendous opportunities for better health and nutrition and protects the environment by reducing soil erosion and pesticide use.' Veneman added, more bluntly: 'With this case, we are fighting for the interests of American agriculture.'

This book should serve as a warning about what happens when economic interests overrule those of the consumer. If the US action succeeds, GM foods will be forced onto European markets regardless of the wishes of consumers. Underpinning the US case is the claim that GM food is safe and healthy to eat. Read this book and make up your own mind.

Acknowledgements

This book grew as a concept over time. People I would like to thank for persuading me to write it include John Stauber in the USA and Michael Gillard and Laurie Flynn in the UK.

I would also like to thank my agent Leslie Gardner for her support and persistence and the editorial staff at Earthscan, including Jonathan Sinclair Wilson and Akan Leander for having the courage to publish the book and see it through the publication process.

I would like to thank all the people who gave up their precious time to be interviewed. Many people who cannot be named for various reasons have helped with information, ideas and advice. They know who they are, but for political reasons and their careers it is better that they remain anonymous. All quotes are from interviews with the author, unless otehrwise stated.

Some people have offered assistance and help beyond what might be deemed reasonably acceptable. A special thanks has to go to the staff at GeneWatch UK – Sue Mayer, Becky Price, Helen Wallace and Rob Johnson – for helping with relentless and often tedious queries on GM. Jonathan Matthews at NGIN also provided invaluable advice and information.

I also want to thank Arpad Pusztai and Susan Bardocz for their hospitality in Aberdeen, and Pusztai for patiently answering so many questions. I would also like to thank specifically Professor Philip James and the staff of The Royal Society for also answering numerous questions and clarifying crucial issues for me.

This book would not have been possible without a small grant of around £3000 from the Network for Social Change and a £1000 grant from the Goldsmith Foundation.

Acronyms and Abbreviations

ACNFP	Advisory Committee on Novel Foods and Processes
ACRE	Advisory Committee on Releases to the Environment
AEBC	Agriculture and Environment Biotechnology Commission
BBSRC	Biotechnology and Biological Sciences Research Council
BIO	Biotechnology Industry Organization (USA)
BMJ	*British Medical Journal*
BSAG	Blairingone and Saline Action Group
BSE	bovine spongiform encephalopathy
CaMV	cauliflower mosaic virus
CAP	Common Agricultural Policy (EU)
CAUT	Canadian Association of University Teachers
CEI	Competitive Enterprise Institute (USA)
CFFAR	Centre for Food and Agricultural Research
CJD	Creutzfeldt-Jakob Disease
CMO	Chief Medical Officer
CSA	Chief Scientific Advisor
CVL	Central Veterinary Laboratory
Defra	Department for Environment, Food and Rural Affairs
DoH	Department of Health
DTI	Department of Trade and Industry
EEDA	East of England Development Agency
EPA	Environmental Protection Agency
ESEF	European Science and Environment Forum
EU	European Union
FAO	Food and Agriculture Organization (UN)
FCCI	Food Chain and Crops for Industry
FDA	Food and Drug Administration (USA)
FSA	Food Standards Agency (UK)
GE	genetically engineered
GM	genetically modified
GMO	genetically modified organism
HEFC	Higher Education Funding Councils

HSE	Health and Safety Executive
IARC	Institute of Arable Crop Research
IEA	Institute of Economic Affairs
iPCR	inverse polymerase chain reaction
IPMS	Institution of Professionals, Managers and Specialists
IPN	International Policy Network
JIC	John Innes Centre
MAFF	Ministry of Agriculture, Fisheries and Food
MBM	meat and bone meal
MLC	Meat and Livestock Commission
MRC	Medical Research Council
MRM	mechanically recovered meat
NAFTA	North American Free Trade Agreement
NFBG	National Federation of Badger Groups
NHS	National Health Service
NFU	National Farmers' Union
NPU	Neuropathogenesis Unit (Edinburgh)
NRC	National Research Council (USA)
OECD	Organisation for Economic Cooperation and Development
OIE	Office International des Epizooties
OPs	organophosphates
OST	Office of Science and Technology
OTM	over thirty months
OWL report	Strategic Review of Organic Waste Spread on Land
PANNA	Pesticide Action Network North America
PCR	polymerase chain reaction
PGST	permanent global summer time
PHLS	Public Health Laboratory Service
PrP	Prion Protein
SAP	Scientific Advisory Panel (of EPA, USA)
SBO	specified bovine offal
SCAN	Scientific Committee for Animal Nutrition (European Commission)
SCRI	Scottish Crop Research Institute
SEAC	Spongiform Encephalopathy Advisory Committee
SEPA	Scottish Environmental Protection Agency
SIRC	Social Issues Research Centre
SMC	Science Media Centre
SVS	State Veterinary Service
TSE	transmissible spongiform encephalopathy
USAID	United States Agency for International Development
USDA	United States Department of Agriculture

vCJD   variant Creutzfeldt-Jakob Disease
WHO   World Health Organization
WFA   Wholesome Food Alliance
WTO   World Trade Organization

CHAPTER 1

Introduction

'In GM crops, I can find no serious evidence of health risks'
Tony Blair, speaking to The Royal Society, May 2002[1]

'Political language is designed to make lies sound truthful'
George Orwell[2]

'Don't worry – it's safe to eat' is a familiar phrase uttered by both politicians and scientists to reassure consumers that the food they eat on a daily basis is safe. It was issued like a mantra throughout the BSE crisis and we are now being told the same about genetically modified (GM) foods.

As vast swathes of the developing world struggle to eat enough food, many in the developed world continue to be worried by what they eat, so much so that the food safety issue remains a hot political topic. Over two-thirds of people within the European Union (EU) are worried about food safety.[3] A recent survey in the UK found that two-thirds of UK consumers describe themselves as 'very' or 'quite' concerned about food safety.

The Food Standards Agency (FSA), the new body responsible for food safety in the UK, who commissioned this latest research, is at pains to point out that, although this number is high, it has dropped slightly over the last two years. So have the number of people who are concerned about BSE (bovine spongiform encephalopathy) – although just under half (some 45 per cent) remained worried about it in the UK.[4]

The FSA will use apparent declining consumer concern to justify relaxing the rules that have been put in place to protect the consumer against BSE. In February 2003, Sir John Krebs, the Chairman of the FSA, indicated that this would happen, a move considered premature by those who represent the families of the 122 people who have died from the human equivalent of the disease (known as variant Creutzfeldt-Jakob Disease or vCJD). As we shall see in Chapter 2, the evidence and the science are not as clear-cut as the government might like us to believe.

A re-occurring phenomenon during the BSE crisis was that critical scientists were marginalized and vilified, and Chapter 3 looks at what happened to some of the prominent scientists and government critics during the BSE crisis.

But BSE is supposedly all behind us now. The British government reassures consumers that the torrid lessons of the BSE saga have forged a new political landscape; one where openness and the consumer reign supreme. Alan Milburn, British Secretary of State for Health, argues that in response to BSE the government has 'sought to put the consumer at the heart of the decision-making on food safety issues and we have established the independent Food Standards Agency. We have opened up our Scientific Advisory Committees and put scientific advice to government in the public domain, encouraging a culture of openness.'[5]

In essence, this book examines that one paragraph, whether the lessons have been learnt, and whether there is a culture of openness surrounding science, food and farming. It is also especially relevant to the debate on genetic engineering. Fundamentally it is whether consumers' interests are really being put first by the government.

Indeed, although the introduction of the FSA has to be seen as a good step forward, its record since its inception does not promote confidence in the protection of the consumer. The role of its Chairman, Sir John Krebs, in promoting GM food and undermining organic produce has been questioned by consumer organizations and is explored in Chapter 10.

It is also the reason why in Chapter 4 this book examines what happened during the foot and mouth crisis in the UK. Whilst you could argue that this is not directly a food safety issue, the response to the crisis shows that, contrary to what the Labour government would like us to believe, nothing much has changed since the dark days of the BSE crisis, despite a change of government. Once again critical voices were marginalized.

The issues of BSE and foot and mouth disease both leave some very difficult questions unanswered, but one major lingering doubt is whether corporate interests or consumer interests will win the debate over food safety. Nowhere will that be more of an issue than with the topic of GM foods.

We stand on the verge of what is called the 'biotech century', in which genetic engineering will alter our world in quite profound ways, from GM foods to 'bio-pharming' (crops that produce vaccines and medicines) to cloning. We are promised that this genetic revolution could end world hunger and eradicate the diseases that continue to plague us.

Whilst politicians and the biotech industry promote GM, consumers in Europe, at least, are hostile to it. Surveys show that 94 per cent of Europeans want the right to choose whether to eat GM food and a staggering 71 per cent do not want to eat it at all.[6] In the UK at the beginning of 2003, consumers were meant to be at the centre of the 'public debate' on GM that will decide whether and how commercialization of GM will happen. The majority of the public in the UK is firmly against commercialization. A survey by the Consumers' Association in September 2002 found that 68 per cent were against the growing of GM crops for commercial purposes and some 57 per cent had concerns about GM foods.[7]

The unpalatable truth for governments across Europe is that people have real worries over GMOs (genetically modified organisms), and that these cannot be as easily dismissed as the pro-GM lobby would like. 'Consumer concerns about GM foods have been all too easily dismissed as emotional, anti-science and anti-progress', said the Consumers' Association.[8]

People worry about corporate control of our food supply. Over 60 per cent of the international food chain is controlled by just ten companies that are involved in seed, fertilizers, pesticides, processing and shipments. Cargill and Archer Daniel Midland control 80 per cent of the world's grain, and Syngenta, DuPont, Monsanto and Aventis account for two-thirds of the pesticide market. Just four companies control the supply of corn, wheat, rice and other commodities.[9]

People worry about the health effects of GM food. In part to allay these fears, in May 2002 Britain's Prime Minister, Tony Blair, told the scientific establishment in the UK, The Royal Society, there was no health risk from GM foods. However, one scientist who believes that he has uncovered critical evidence that GM crops cause harm is Dr Arpad Pusztai, who spoke out on *World in Action* in 1998. The full story of what happened to him is told in Chapter 5. Pusztai was castigated by The Royal Society, who still hound him to this day, and their role is critically analysed in Chapter 6.

In marked contrast to The Royal Society in the UK, The Royal Society of Canada's Panel of Experts into Biotechnology has identified 'serious risks to human health such as the potential for allergens'.[10] The risks associated with GM and a number of flaws in the US and UK regulatory systems are examined in Chapter 7.

But Pusztai is not alone in being vilified for raising concerns about GM. One burning issue surrounding GM is contamination of non-GM and organic food. These fears were fuelled by research published in

Nature that showed that GM had contaminated native maize in November 2001. The scientists who published the research, Ignacio Chapela and David Quist, have also been attacked. Their story is told in Chapter 8.

Part of the reason that Chapela believes he was vilified is that he had opposed a large corporate tie-up between the university where he worked and a huge transnational biotech company. Increasingly, this is becoming the norm, and the issue of the corporatization of science is considered in Chapter 9.

Finally, in Chapter 11, the book looks at radical ways to reform agriculture and science. As we face decisions on GM commercialization our judgement is guided not only by looking to the future, but also to the past. This time we must learn from our mistakes, rather than repeat them.

Over the last 50 years the preoccupation has been to provide cheap food. Whilst this might have made sense decades ago, the ravages of foot and mouth disease and BSE have meant that the cheap food mentality is no longer plausible in the long-term. The most persuasive argument for change is that the current system is neither safe, nor economically or environmentally sustainable. This mentality has produced an inefficient, intensive system in which the world's richest 16 countries subsidize their agricultural sectors to the tune of US$362 billion a year, whilst 2 million children in the developing world die each year because of contaminated water and food supplies.[11]

We know that in the UK the agriculture industry is in a mess. The National Farmers' Union (NFU) talks about a 'decade *horribilis*', in which BSE, foot and mouth disease, exchange rates and commodity prices have all created problems for the UK farmer.[12] Since 1997, farm incomes have fallen by 75 per cent and the average farm income at the time of writing is now less than £10,000. At least one farmer or farm worker commits suicide each week.[13] The average age of a UK farmer is 58, with little indication of an influx from the younger generation.[14] Demand for their products is also declining. Over the last two years alone some 7.5 million British consumers have reduced their consumption of meat.[15]

Moreover, cheap food is actually a myth. Consider the external or hidden costs. It costs nearly £120 million to remove pesticides from drinking water every year. Atmospheric pollution from UK agriculture causes indirect costs of some £1 billion, including those brought about by climate change.[16] 'We pay three times over', says Professor Jules Pretty from the University of Essex. 'Once at the till, then again through tax for EU subsidies and then again through hidden subsidies.' The hidden costs of industrial agriculture – damage to soil structure, pollution and environmental damage – Pretty believes, is in the region of £2.34 billion a year.[17] Our biodiversity is slowly being eroded. Two-thirds of England's hedgerows were lost between the 1950s and 1990s[18] and nearly two-thirds of the population of skylarks has been wiped out in the last 30 years.[19]

Furthermore, our cheap food policy has not produced healthy people. We are eating too much of the wrong food and not enough of the right food. In the USA, where over 60 per cent of people are already overweight or obese, 90 per cent of money spent on food is spent on processed food. French fries are now the most widely sold food item in the USA and each person already drinks over a gallon of soda a week, while some teenage boys drink five or more cans per day.[20]

In the UK, diets are not much better and are getting worse. People are eating more and more processed and packaged food, and less and less fresh food. The demand for frozen convenience meat products has grown by 127 per cent since 1978.[21] People are literally losing the ability to cook. At least three-quarters of UK men and woman are expected to be overweight within 10 to 15 years, and obesity could become as important a health issue as smoking.[22]

The diet of children in the UK is similarly appalling, with over £8 million spent each week on junk food. On average, children in the UK eat less than half the recommended amount of fruit and vegetables; and between 75 and 93 per cent eat more saturated fat, sugar and salt than is recommended for adults.[23] Of the food adverts shown during children's TV, 99 per cent are for products that are high in fat, sugar or salt.[24] UK childhood obesity rates are reaching epidemic proportions, with one in five nine-year olds overweight and one in ten obese – rates that have doubled in the last ten years.[25]

This Western cheap food policy has not produced appetizing food. The current supermarket status quo is summarized by the food writer Joanna Blythmann as 'permanent global summer time' (PGST), a 'curiously uniform, nature-defying new order' where tasteless, unvarying imports take over from fresh, locally grown produce.[26] Similarly, this policy has not produced safe food. In the USA, 76 million people suffer from food-borne diseases every year, leading to some 5000 deaths.[27] In the UK, it is estimated that 4.5 million people suffer from food poisoning every year.[28]

What the cheap food policy has produced is a system that is controlled by just a handful of companies. If this corporate domination continues it will lead to a future where livestock are reared on a massive scale – they never lead a natural life, never see a blade of grass, or never see natural light. Such places are already being proposed. One Dutch proposal is for a farm that would be a factory occupying six floors and covering 200 hectares. Everything would be grown or reared inside, including 300,000 pigs, over 1 million chickens, and tens of thousands of fish and vegetables.[29]

The result would be a food supply in which biotechnology is the norm. Milk, butter and meat from cloned animals may become available in the UK in 2003 from the USA, after US scientists deemed that cloned produce does not pose a food safety concern.[30] Despite the fact that Dolly, the first cloned animal, died at a worryingly premature age, we are told that GM animals are like GM food – that is, safe to eat. The pro-biotech lobby will tell you that the technology is safe, proven, has environmental benefits and is necessary for feeding the world. They argue that if no one has died or fallen ill there is nothing to make a fuss about.[31]

The use of GM crops is expanding rapidly and the estimated global area of GM crops for 2002 is 58.7 million hectares grown by 6 million farmers in 16 countries. In 1996, only 1.7 million hectares were grown. Four countries continue to produce 99 per cent of the world's GM crops: the USA, Argentina, Canada and China.[32]

Whilst proponents argue that this technology will feed the world, the majority of crops so far planted have been engineered to be tolerant to herbicides, guaranteeing the continuation of our chemical age. One herbicide still dominates: Monsanto's Roundup. Monsanto is at the forefront of GM crop production and it would appear to be a cunning tactic for a company to produce a GM plant that is resistant to its own herbicide. Senior Monsanto executives admit that the cornerstone of its business is Roundup, and that biotechnology is just the delivery vehicle.[33] 'The company wants farmers to use lots of glyphosate – the active ingredient in Roundup – over lots of acreage. The idea is to try to dominate the seed and herbicide market as one system,' says one North Dakota farmer.[34]

Dr Guy Hatchard, an ex-Director of Genetic ID, a leading tester of GM crops, says simply: 'A [US]$3 billion international biotech-herbicide company has worked out how to manipulate the trillion-dollar US agrifood industry. This practice has made no financial sense for farmers.'[35]

Hatchard's concerns were borne out by research by the UK Soil Association that found that GM crops have been an 'economic disaster' in the USA and Canada, costing the US economy US$12 billion in farm subsidies, lower crop prices, loss of major export orders and product recalls.[36] Even the United States Department of Agriculture (USDA), a prominent backer of GM, concluded that after looking at farm incomes in its own study: 'perhaps the biggest issue raised by these results is how to explain the rapid adoption of GE [genetically engineered] crops when farm financial impacts appear to be mixed or even negative.'[37]

The other main use of GM plants has been pest control, mainly by inserting a natural insecticide into a plant. The only currently commercial insect resistant crops are *Bt* crops, containing a gene from the soil bacterium *Bacillus thuringiensis*. But patents taken out by Monsanto state that transgenic pest control 'may not be desirable in the long term' because it produces resistant strains and 'numerous problems remain ... under actual field conditions'.[38] As we shall discover in Chapter 7, there are growing health concerns related to *Bt* crops.

As US farmers flounder back home, the push is on internationally, and their lobbying is a multi-million pound effort. Take, for example, the Biotechnology Industry Organization (BIO) in the USA, which nine years ago had a budget of US$1.8 million, but by 2002 employed over 70 staff, represents 1000 companies and has a budget of US$30 million.[39]

There is currently a moratorium on GM crops across Europe. This is seen as illegal by the biotech industry and the US administration, who believe it could 'halt world trade in biotech products'.[40] There is US pressure for the UK to start commercial planting, with the Washington press urging the USA to take Europe to the World Trade Organization (WTO) and the USA looking like it will.[41]

With Europe currently cut off, the biotech industry is trying to promote the technology across swathes of the 'developing world'. Part of their effort involves persuading people that their underlying reason is basically philanthropic – that is, to feed the world. But food experts do not believe them. The Chair of the International Obesity Taskforce is Professor Philip James who says: 'The idea that GM is suddenly going to solve the world's hunger is such crass naivety.'

The PR tactics entered a sinister phase during the famine in southern Africa in 2002. Just as Europeans were being pressured to take GM food, so too was pressure put on six countries in southern Africa to accept GM food aid, whilst conventional food aid was not offered.[42] At the height of the furore, the World Food Programme admitted that it had been delivering GM food as emergency aid for seven years without informing the countries concerned.[43]

But it was the tactics and language of the US government and pro-GM scientists that alarmed many. Suddenly GM became 'life-saving technology',[44] whereas those who stood in the way of the biotech industry were condemned as murderers. 'The Bush administration is not going to sit there and let these groups kill millions of poor people in southern Africa through their ideological campaign', said Andrew Natsios of the United States Agency for International Development (USAID).[45] Despite this, environmentalists argued that there were numerous sources of non-GM aid available around the world, including in the USA.[46]

As GM-critics are accused of killing millions, leaked PR documents show that the touchstone issues of the 'environmental and human health risk' of GM foods are seen as 'communications killing fields' for the biotech industry in Europe. They are not to be discussed, and 'the industry must accept that it is for those charged with the public trust in this area – politicians and regulators – to assure the public that bio-industry products are safe'.[47]

This is precisely what Tony Blair was trying to do when he told The Royal Society that with 'GM crops I can find no serious evidence of health risks'. The sentence is meant to close the debate about health and GM. But how does one define 'serious', and has Blair even looked? Time and again, absence of evidence is given as evidence of absence, but that argument is no longer acceptable as we formulate a truly sustainable food and farming policy that does put consumers first.

In essence, to say there are no health risks with GM is like saying a decade ago that eating UK beef was safe. The Official Inquiry into BSE noted that 'no one foresaw' the problem of recycling animal protein.[48] Essentially, 50 years of the safe use of recycling animal protein was not a safeguard of public health. Something unexpected happened that caught everyone unaware. This is exactly what critics of the GM industry believe will or could happen with GMOs.

But, as you will see, the stranglehold of silence is slowly tightening on the debate over genetic engineering, as short-term economic needs are, once again, seeing precaution thrown to the wind. Our chance as consumers of understanding the risks involved in eating GMOs is slowly being eroded. Every day it gets harder to weigh up the risks when you are told: 'Don't worry – it is safe to eat.'

CHAPTER 2

Safe to Eat

'For ten years the government told the people:
– There is no evidence that BSE can be transmitted to humans;
– It is most unlikely that BSE poses any risk to humans; and
– It is safe to eat beef'

The BSE Inquiry[1]

Why – Why Me?
Help me, I'm tired
Trapped in myself, I have to get out
Why me, what did I do
Can't you hear me struggling, can you hear me shout
Looking for answers and what made me this way
Feel so helpless, got to live another day
Feeling the rage, inside it builds so fast
Got to strike out – How long will it last

Peter Hall, vCJD victim: 17/2/75–8/2/96[2]

BSE – a distant memory or living nightmare?

It is July 2002 and a typical summer's day in London, when the rain will get the better of the sunshine. Today's headlines report that an American bomb has killed 30 guests at an Afghan wedding and how a 108 year-old pensioner named Alice has starved herself to death due to the closure of her nursing home.

Unreported is a meeting organized by the FSA at the Congress Centre in Great Russell Street, just a stone's throw from the hustle and bustle of Tottenham Court Road. It is seen by some as a fundamentally important meeting on food safety, as one of the pillars of health protection in relation to BSE is under review. But no one from the public is here.

Inside, almost 120 delegates, mainly from the meat industry, have gathered to discuss the FSA's proposed review of its 'Over Thirty Months (OTM) Rule', whereby all cattle in the UK are slaughtered at the age of 30 months as a precautionary measure to protect against BSE. The disease that brought UK agriculture to its knees in the 1990s, BSE is now seen as a distant memory, one that has been replaced by more recent disasters to afflict UK agriculture, such as foot and mouth.

The fact that the UK government still has protective measures in place against BSE shows that the problem continues to pose a threat to health more than people realize. The delegates at the OTM meeting talk about a new dawn for UK agriculture, but the long shadow of BSE still haunts them.

Others are worried too. The disease's niggling refusal to do what the scientists have predicted baffles them. Some scientists now talk about 'the third phase of the BSE epidemic',[3] a much, much smaller phase than the original one, but what is annoying about it is that it should not be happening at all.

Both BSE and its human equivalent, Creutzfeldt-Jakob Disease (CJD), are what is known as TSEs or 'transmissible spongiform encephalopathies', fatal diseases that destroy the nervous systems and ultimately the brains of their victims, where the disease produces characteristic spongy holes. By mid-1987, it was known that the disease could cross the species barrier.[4] Scrapie was found in sheep and goats, transmissible mink encephalopathies in mink, and 'kuru' in humans, a disease that inflicted tribes in Papua New Guinea that had resorted to cannibalism. Rodents could also be infected.

To the list of TSEs would be added BSE – bovine spongiform encephalopathy – which would commonly be dubbed 'Mad Cow Disease', after the vivid pictures of staggering, stumbling cows. TSEs were called transmissible because the disease can be transmitted; spongiform due to the spongy holes in the affected brains, and encephalopathies from two Greek words, enkephaloo, for brain, and pathos, for suffering.

For all the diseases, which are characterized by their hardiness and the difficulty with which they are inactivated, there is no cure and the incubation period is normally years in smaller animals and decades in humans. There then follows a short illness, which inevitably leads to a harrowing and painful death.

Ironically, one of the fundamental flaws in the government response to the BSE crisis was to believe that a transmissible disease like BSE – that was known to cross many species barriers – was actually non-transmissible from cows to man. The assumption was that cows were in fact a 'dead-end host'. This was described as the 'big lie' by one of the key critics of the government's handling of BSE, Professor Richard Lacey. The reason for this was that some scientists thought BSE was scrapie in cows and if we had lived with scrapie in sheep for over 200 years without humans becoming infected, the reasoning went that we could live with scrapie in cows. The problem was that no one asked the true experts, who were shunned, and no one did any research to check the theory.

The lies would underpin the reassurances that would turn out to be false. During the crisis, consumers were repeatedly told that UK beef was safe to eat. But this was a 'certainty' that politicians knew to be uncertain.[5] This political reassurance was typified by the then Agriculture Minister, John Gummer, trying to feed his daughter a hamburger in front of the media. Unfortunately it was too hot and Gummer had to eat the burger himself.

Whilst the politicians uttered public words of reassurance, many scientists privately stopped their families eating beef. 'I was there at the first meeting of the directors of research when we were quietly briefed in 1986', says Professor Philip James, one of the UK's leading experts on food safety, who believes the BSE crisis could have been avoided. 'It was obvious that as we discussed the science there was enormous uncertainty. I remember going back and phoning my whole family telling them not to eat any beef products.'

Although experts believe the disease is difficult to transmit between species, this does not mean that transmission does not occur. 'Those who I regard as my peers in this subject,' one senior TSE scientist told me, 'from the first time that BSE was known about – have always, always regarded that transmission to humans is a possibility. I know nobody who I would count in any sense a good expert or even a moderate expert, who would have said that it can't be transmitted to humans.'

But the 'it can't be transmitted to humans' lie would permeate publicly for ten years until 20 March 1996, when Stephen Dorrell, the then Conservative Secretary of State for Health, told MPs something that had been denied for a decade. Ten young people had contracted a new variant of 'the harrowing, and invariably fatal, Creutzfeldt-Jakob Disease – vCJD – and that it was probable that they had caught BSE'. He read out a statement from his scientific advisors, the Spongiform Encephalopathy Advisory Committee (SEAC), that exploded a political time bomb. 'On current data and in the absence of any credible alternative, the most likely explanation at present is that these cases are linked to exposure to BSE,' said Dorrell.[6]

Those words destroyed people's trust in politicians to tell them the truth about food safety. People felt 'betrayed'. According to Dr Erik

Millstone, a World Health Organization (WHO) consultant on BSE, it was confirmation 'of the biggest single failure in UK public policy since the Suez crisis in 1956'. One SEAC scientist said it was 'a fucking disaster'.[7]

'I think it came as a terrible shock, the timing of the first human cases, in 1994–1995,' says Professor Lacey now, 'because they thought they would get away with it for another period of government. That's where they went wrong'.

The political and economic fallout from Dorrell's statement was immense. The press accused politicians of 'gambling with the health of the nation' and 'sacrificing the population'. The British Meat Federation still call 20 March 1996 a 'fateful day' when they faced a 'total crisis'. The UK meat industry's export markets collapsed 'like a house cards'. Ever since they have tried to regain 'free access to the EU and world markets'.[8] The announcement also led to the setting up of the OTM Rule, which persists to this day to protect the consumer against BSE. But does it?

Relaxing the rules?

Eight years after Dorrell's announcement there is intense pressure from the beef industry to relax the OTM Rule. For UK beef to be exported again, it needs to be seen as safe. If it is safe, then the rule can be relaxed. If it is not safe, then the rule should stay. But there is a stark contrast between a measure seen by the government as a pillar of food safety and considered by the farming industry as nothing more than a public relations exercise.

The FSA sees the OTM Rule as 'one of the key BSE controls introduced to protect food safety'.[9] But senior members of the farming community believe that the rule 'is not a BSE control measure'; it was introduced to 'restore consumer confidence' in UK beef.[10]

When the rule was introduced 'no one knew, or could reliably calculate, how many cattle sub-clinically infected with BSE were entering the food chain'.[11] Now we are told that under 30 months beef is safe and over that period it is unsafe. When scientists first considered a time limit on cattle entering the food chain, they thought that 24 months was more appropriate. Professor Peter Smith, the current chair of SEAC, who will look at the risks of relaxing the rule, argues that the 30-month cut-off is based on science. The risks associated with the consumption of OTM cattle were estimated to be several hundred-fold greater than those associated with under 30-month old cattle, he says.[12]

Government critics argue that the OTM rule is a classic fudge, merging politics and health, that reduces but does not eliminate the risk of infection. Dr John Godfrey from the consumer organization Foodaware believes the answer lies in the fact that by 30 months most beef is ready for market.[13] 'The OTM Rule does not eradicate the risk of BSE, it only diminishes it' adds Dr Erik Millstone. 'No one really knows, however, by how much it is diminished, but it is probably quite significant. But it is not a guarantee that you won't be exposed to the BSE pathogen.' Professor Richard Lacey goes further, arguing that the idea that cattle are safe to eat up to 30 months is pure 'nonsense'.[14]

Whichever way you look at it, over 5 million cattle have been slaughtered under the OTM Rule.[15] The government sees lives saved; the farming industry sees lost profit. The meat industry has been trying to rid itself of the ban. It walks a tightrope between selling more beef by relaxing the rule, and selling beef that contains BSE. If that happens the industry would face an irreversible predicament. Society cannot withstand another BSE crisis. We do not even know the financial or personal cost of the current one.

However, there are real reasons for at least changing the ban. Mrs Ena McNeil is the branch Chairperson for the North Uist Scottish Crofting Foundation. North Uist is a remote and beautiful island in the Outer Hebrides. Mrs McNeil feels so strongly about the OTM Rule she travelled to the FSA meeting – over a day's journey. Her son farms the beautiful Ardbhan fold of highland cattle, a shaggy horned cattle of distinctive character and hardiness. A quarter of the McNeil's flock is the rare black Highlander. The whole herd is traditionally reared organically and extensively on pasture. The problem is that these traditional breeds take so long to reach maturity – traditionally about four years – that they are killed well before their beef will be at its best. In fact, because of the OTM Rule, they are killed when the bullocks are immature. This means that the farm loses money, and what was already a marginal income becomes critical.

There has only ever been one case of BSE in the Outer Hebrides, and that cow came from the mainland. BSE is as removed from the highland cattle as the streets of London are from North Uist, but Mrs McNeil and her family continue to be punished financially for a problem that had nothing to do with them.

Mrs McNeil listened to the speakers agreeing whole-heartedly with the view of Jim Walker, the President of the NFU Scotland. Walker told the audience that the UK's self-sufficiency in beef has declined from 74 per cent in 1999 to about 54 per cent in 2003. The NFU in Scotland believes that the OTM Rule needs to 'unravel' and is calling for all cattle over 30 months to be tested for BSE.

Those animals that are BSE-free and were born after the 1 October 1998, when the farming industry introduced its cattle passport scheme, should be re-admitted to the food chain, argued the Scottish NFU. Walker also advocated that there needs to be a 'regional' approach to the review, wary that many of his members will have livestock, like the McNeil's cattle, from slower maturing breeds.[16] But the OTM conference went eerily quiet when Lester Firkins from the Human BSE Foundation stood up to speak. The Foundation is only an observer in the FSA's review, not an active stakeholder. Firkins stood tall and proud and told the audience he represented the families who have lost relatives due to vCJD. He lost his son from vCJD the year before. Although he welcomed the review, he wanted to bring some balance to it. He did not like the tone of the Scottish NFU that the OTM scheme should 'unwind'. There could not be any foregone conclusions. 'Consumers', he said, 'have to know the risks they are taking.'[17]

This is the conundrum consumers are still left with. The UK farming industry, the meat industry, the supermarkets, the retailers, the politicians all want the scientists to tell the consumer how safe it is to eat UK beef. The very unpalatable truth is that, although the scientists can guess, they still do not know for sure. 'We have pleaded with the scientists to give us definitive answers, but we are never going to get them,' says Jim Walker.[18]

Indeed, can science ever prove safety? As the scientists search for answers, the FSA has promised that 'consumer safety' would be the overriding concern of the OTM Rule review. The FSA is one of the positive outcomes of the BSE saga, a supposedly independent agency that now fights for the consumer. However, as we shall see in Chapter 10, it has already made certain decisions on organic food and GMOs that many believe have gone against consumer interests; but what will it do on BSE?

The consumer advocate at the conference, Dr John Godfrey, argued that the OTM Rule is an important safeguard against vCJD. 'The time has not yet come to scrap it,' he told the conference.[19]

Speaking two months after the OTM meeting, Lester Firkins said that he felt the FSA 'will have to be seen to give something' and might allow OTM beef back on to the market if there is adequate testing. 'I think I am building up quite a case to say that this testing business is not enough,' he says. As if to prove his point, that same day the European Commission released a report showing that abattoir staff were still failing to observe basic anti-BSE measures 16 years after the crisis had started.

In February 2003, the head of the FSA, Sir John Krebs, hinted that the OTM Rule would soon be relaxed. 'Thirty months is not a magic number', he said.[20]

The only certainty is uncertainty

Sixteen years after the disease first struck, there is a huge amount that we still do not understand about BSE. We still really do not know what started the BSE epidemic or how safe it is to eat UK beef.

For a scientific area that has generated two Nobel Prize winners, the area of TSE research remains a minefield of competing scientific camps and a hotbed of controversy. We know that all TSEs are essentially one disease that occurs in different species, with many different strains of agent, just as flu has different strains. We do know that BSE and vCJD are both forms of TSE. We also know that the strain of the infectious disease that causes BSE has identical properties to the one causing vCJD.[21] But we don't know, for example, whether the infectious agent that causes BSE and vCJD is more easily transmitted to humans than other strains of TSEs. Nor do we know for sure what the infective agent is or even how it works, whether it is a rogue protein called a prion or slow virus, or something called a virino. Whilst PrPs – prion proteins – have been accepted by many researchers as being involved in the infection process, we still don't really know what they are and whether they are the infectious agent, or only part of it. There is consensus, though, that PrPs are unique and pretty indestructible, and withstand heating, burying, boiling, and radiation.

'It is still not clear whether the prion is the infective agent', argues Dr Erik Millstone. 'It is not clear whether the prion is the pathogenic agent or whether it accumulates as a consequence of the disease process. We also do not know whether one pathogenic particle is sufficient to trigger an infection. We do not know if there is a safe threshold.'

We also do not know if eating is the main infection route. A more direct route may be through scarification – cuts and grazes. 'I am convinced that BSE and CJD are the same but I am not convinced that it is all acquired through ingestion. There is no evidence [for] and a lot of evidence against that the infection is acquired by eating,' says Professor Lacey. 'It could be through cuts, inhalation and grazes.'

Additionally, we don't know how many cows are infected. Most of the analysis has been done via computer modelling and risk assessments. There are those who argue that computer models are only as good as the input data. 'Risk assessment is important,' says Professor Colchester, a consultant neurologist, who has played an active part in the BSE saga. 'The negatives are … that it is a classic trap. Science that is based on certain assumptions may look impressive and you get a fat report with loads of figures, but of course all the arithmetic is no stronger than the input assumptions. An awful lot of people do not understand how limited these risk assessments are.'

We do know how many official cases of BSE there have been, although we will never know the number of cows that died in the pre-clinical stages of the disease. As of the end of March 2002, there had been 182,552 cases of BSE recorded in the UK.[22] Recent statistical analysis suggests that the original epidemic may have been twice the size first thought, with some 1.6 million infected animals eaten out of a total of 2 million infected.[23] At the peak of the epidemic approximately 200,000 infected animals a year went into the food chain, including some 10,000 late stage BSE animals, known to be the most infectious.

The bad news is that we are still recording new cases of the disease, some 781 in 2001, and 243 in the first six months of 2002. The good news is that the number is declining by about 40 per cent a year.[24] But those are the official figures. We are not sure how many cows slaughtered over 30 months harboured the disease.

We also know that a number of cattle continue to come down with BSE, cases that scientists believe defy all explanation – the new third phase of the disease. This total includes 21 cases born after mid-1996, when all potential routes of infection should have been stopped. Although it is a small number, it is a hugely worrying statistic. Scientists do not know how these cattle became infected. There is, according to Peter Smith, 'no convincing explanation found for this in epidemiological terms'.[25] It could mean that there is a background level of BSE in the cattle population. It could even mean that there are other factors at work that facilitate BSE.

Some analysts, like Professor Richard Lacey and another government critic Dr Harash Narang, argue that they can explain these 21 cases by vertical transfer – that is, that the disease was passed from mother to calf. 'It is vertical transfer,' says Lacey. 'They put the feed ban in place in July 1988, well it is now 14 years on from that. I don't believe it wasn't adopted thoroughly. It is another lie.'

'They don't want to take note of all the work that has been done on sheep scrapie,' continues Professor Lacey. 'There is absolutely unquestionable evidence that it can be acquired from the environment and by vertical transfer.' If Lacey is right, an unknown number of cows are passing the disease to their young. The parent may then be killed or eaten before we realize it had BSE or that it had passed it on. It means that we might have BSE for decades.

One world-renowned expert on scrapie, who has evidence that environmental or pastural contamination could occur, is Dr Alan Dickinson. In one experiment in the 1970s, Dr Dickinson took an uninfected mouse and fed it with a mixture containing brain scrapie. One of his researchers sat up all night to collect all the mice droppings. 'We looked at the infectivity in those pellets and all the infectivity had gone straight through. There was a peak on infectivity – a very, very high proportion', he says.

What this shows is that healthy animals excrete TSE infectivity. 'But this does not happen with infected mice, they do not excrete infectivity,' says Dickinson. 'What I assume is that until the opposite can be proven, every bit of infectivity fed to any mammal, I guess any bird, will be excreted.' Dickinson does not want to overstate the risk of pastural contamination but believes it could extend the life of the epidemic, exactly what is currently happening.

Although Dickinson's research was never published, rumours of the experiment had reached the Ministry of Agriculture, Fisheries and Food (MAFF) by the late 1980s. One 'Commercial in Confidence' document noted that there had 'been rumours of scrapie-like agents passing through the gut and retaining infectivity'.[26] The authorities knew of the problem, it is just that they never seemed to take any notice.

We know that other countries have BSE too. There have been approximately 2700 cases in ten other European countries. More worrying is that in Germany, Spain, Denmark and Italy there were no recorded cases of BSE before 2000. But there have been since, and in all four countries BSE is on the increase. Another worrying statistic is that in 2001 in nine European countries nearly 300 cattle destined for the human food chain were found to have BSE.[27] In Ireland, although incidence remains low, there were 200 positive BSE cases in 2001, and the numbers have risen every year since 1995.[28]

There are other worrying signs. In late 2001, Japan recorded its first case of BSE, which could be ominous for parts of the south east Asian region where cow brain is eaten as a delicacy. A year later, the Czech Republic experienced its first case of BSE. Experts at WHO now believe that BSE 'has joined AIDS as a major health challenge facing the world'.[29]

There are now worries that, having exported the problem, it is now being re-imported, with infected parts of animals making their way onto our plates. Although there are routine checks to prevent this, in September 2002, the FSA announced that banned spinal cord had been found in frozen beef imported from France and Germany.[30]

It is generally agreed that there needs to be a reliable test for BSE, but we still do not have one for live animals. The only really reliable test is one performed when the animal is dead. So there is still no way of telling if an animal is in the early stages of BSE infection, according to Peter Smith. The problem, says Smith, is that although tests in animals with clinical signs of BSE have now been developed and are deemed reliable, that is not the case for younger cattle that could be in the pre-clinical stage of the disease. The further you go back in time – that is, the younger the cow – the more likely it becomes that these tests 'won't pick up infectivity'. Between the ages of six months and a year, when we know that certain cattle tissue can be infected, it is 'highly unlikely that [tests] will pick up all the animals that are infected'.[31]

Indeed, while Smith points out that the risk from BSE in the food chain remains highest from older animals we 'have to be pessimistic that there will ever be a test that will pick up the infectivity at all the stages of the incubation period, but we may be wrong'.[32]

We do not know how many people are incubating vCJD. We do know it remains a rare disease. In the autumn of 2002, scientists started studying archive tonsil tissue to see how many contained vCJD in an attempt to shed some light on this problem. Their current estimate is that some 120 per million could be incubating the disease, but the figure could be so widely inaccurate that it is meaningless. 'There is still so much we do not know about variant CJD', concedes Sir Liam Donaldson, the Chief Medical Officer.[33]

In truth, we do not know how many people are going to die. Up to February 2003, some 122 families have had their lives ripped apart by someone dying from the human form of BSE, called vCJD. Another eight families are nursing a relative dying of vCJD.[34]

'It is difficult to predict whether the epidemic has peaked,' says Professor Smith, adding that the number likely to die could be 'a few thousand or even tens of thousands, rather than the hundreds of thousands' once feared.[35] New research published in 2002 put the number of deaths at around 200.[36] Other scientists believe it could be more like 1000. But we just do not know, and as one of Peter Smith's predecessors at SEAC noted: 'We are the experiment'.[37]

But despite what the modellers predict, we can also only guess at the incubation periods, which are dependent on three things: the route of infection, the particular strain of TSE agent and the dose. 'Everything in TSE work is dose dependent', says Dr Alan Dickinson. As we have no way of knowing what doses of contaminated beef people ate, the models are really only educated guesswork; they could be wildly wrong. 'The answer is we haven't a clue,' says Professor Philip James.

Once again the international picture is a cause for concern. In the summer of 2002, Canada suffered its first case of vCJD – and 70 patients were told that they had subsequently been treated by the same surgical instrument with risks of contamination.

Moreover as Dr John Godfrey pointed out at the July seminar, just because the 'statistics do not show cases are rising, this does not exclude that the cases may rise'. The genotype of an individual affects the nature of the disease. So far people struck down with vCJD are mainly from one of three genotypes. The other two may have longer incubation periods. Another argument is that the other two genotypes may be resistant, but could transmit the disease to others who are susceptible.[38]

We do not know if BSE is restricted to cattle. There are serious worries now about BSE in sheep. Many experts believe that, because sheep ate the same meat and bone meal believed to be the cause of BSE, they may also have become infected with BSE. 'Since BSE may be transmissible from sheep to sheep, getting rid of the meat and bone matter might not get rid of BSE in sheep', says Rowland Kao of Oxford University, who examined the issue in 2001. 'That's the thing that really concerns everyone.'[39]

We know that experiments have shown that it is possible to infect sheep with BSE by feeding or injecting them with brain tissue from cattle infected by BSE. However, no sheep have actually ever been found to have contracted BSE 'naturally' this way, so the risk is seen to be theoretical.[40]

Some experts are worried that scrapie could be masking BSE in sheep and that many cases go unreported. If there were BSE in sheep the human population is routinely exposed to large doses of the animal's central nervous system by eating lamb chops, argues Dr Alan Colchester. 'We are not being prevented from eating central nervous tissue'. Colchester believes that mutton and lamb chops could constitute the highest risk of continuing BSE infection in our diets.

We should have known whether sheep harbour BSE by now, but a four-year study to examine whether the UK sheep flock was infected ended in complete farce in 2001 when the government announced that experiments had been performed on cattle brains by mistake.[41]

There are so many 'don't knows' surrounding BSE, it is surprising that anyone can say anything with any authority. If and when the OTM Rule is relaxed, it is to be hoped that the decision makers get it right, both for the families of vCJD victims, for us all as consumers and for farmers like Ena McNeil. Because if the experts and politicians get it wrong again the consequences are too awful even to think about.

The first BSE outbreak

Although the origin of BSE will probably never be known for certain, the most widely accepted explanation is that it was caused by making carnivores out of herbivores.[42] Cows were fed what is called MBM, or meat and bone meal, which was essentially the waste from the carcass – the skull, brain, guts and bone and any remaining scraps of meat. This was done as a way of giving cows extra protein and the rendering industry extra profit. It was also the reason why BSE became more prevalent in dairy herds than beef herds, the latter often being more extensively farmed.

The government had been warned by the Royal Commission on Environmental Pollution as far back as 1979 about the risks of processed animal waste. 'The major problem encountered in this recycling process is the risk of transmitting disease-bearing pathogens to stock and thence to humans', concluded the report.[43] Yet within months, MAFF was urging deregulation and that the rendering industry should 'determine how best to produce a high quality product'. Some have argued that it was deregulation that led to the beginning of the epidemic, although this was dismissed by the official inquiry into BSE, chaired by Lord Phillips, which noted that 'rendering methods have never been capable of completely inactivating TSEs'.[44]

So even now – some 20 years later – we are still unclear how BSE got into cows in the first place. One theory is that it was a sporadic outburst, another that it was a cattle strain of scrapie. Indeed there are claims that BSE was not even a new disease, it was a rare disease called 'stoddy' in Yorkshire, but that incidence was very low.[45] But there are no 'definitive answers' as to why the epidemic has been predominantly in the UK, although other countries have numerous cases of BSE. One answer could lie in the fact that the UK was one of the first countries to feed MBM to young calves.[46]

Once the MBM became contaminated with the infective agent, it was only a matter of time before a serious, unseen problem was building up. 'What is clear, it is the recycling of MBM that caused the epidemic', says Professor Peter Smith from SEAC.[47] Or put another way, as Lord Phillips concluded 'BSE developed into an epidemic as a consequence of an intensive farming practice.'[48] Its international spread was due to pure greed. Even when it was suspected in the UK that MBM was the vector for the disease, 25,000 tonnes of contaminated feed were exported to Europe.[49]

The first sign of the current epidemic occurred in late 1984 and early 1985 in the south east of England. Just before Christmas in 1984, a vet in Sussex, David Bee, visited Pitsham farm near Midhurst in West Sussex where the cows were showing 'unusual symptoms'. Bee said this new disease was 'really spooky', something he had never seen before.[50] In February 1985, the first cow from the herd died after developing head tremors, and by the end of April a further five had died.[51]

Then, also in April, a Holstein/Friesian dairy cow called Jonquil on a Kent farm started acting strangely. Over the following year, other cattle too would show signs of aggression and lack of coordination. The local vet at the time, Colin Whitaker, said 'I had no inkling then of the potential scale of it. If someone had told me then of the panic that would be caused, or that there would be 150,000 cases of BSE in 10 years' time, I would have thought they were mad.'[52]

By now the head of one of the cows identified by Bee had been forwarded to a local MAFF laboratory and, from there, onto Carol Richardson, a pathologist at the Central Veterinary Laboratory (CVL). In September 1985, Richardson undertook a post mortem of one of the Friesian cows, known simply as 'Cow 142'. Notes taken show that she recorded spongiform encephalopathy in the brain tissue. Although a colleague agreed with her, a senior neuropathologist at the Laboratory attributed the cow's death to poisoning.[53]

Richardson wrote her report on 19 September 1985, a 'nice sunny day', when she looked into a microscope and saw tiny holes in stained sections of the cow brain. She had seen the same feature many times before in sheep, but never in cows. 'This was the first time I had seen these lesions in a cow', she said.[54] In evidence to the Phillips inquiry, Richardson said: 'this is scrapie in a cow.'[55] It was not until March 1987 that Richardson's findings were realized to be BSE in cattle.

If the government had acted on the findings of Carol Richardson, Professor Roy Anderson, an epidemiologist and member of SEAC, claimed, in February 1998, that one in three or 60,000 cases of BSE could have been prevented.[56]

This was one of the first of many delays that would typify the government's response. Whether this inaction was more to do with cock-up or conspiracy has been a hotly debated topic. The Phillips inquiry was dismissive of Richardson's findings, arguing that 'while we think that she may well have noticed a pathological similarity to scrapie, we do not believe that she can have concluded that Cow 142 was suffering from a transmissible spongiform encephalopathy'.[57] The Phillips inquiry tried to imply that Richardson's memory was at fault. The Richardson case is one where the Phillips inquiry seems to have muddied the waters of history in favour of the then Ministry of Agriculture. The failure to act on what Richardson had seen was 'not a matter for criticism'.[58] Quite why the inquiry was so keen to discredit Richardson has puzzled many people. It has been argued that the later the disease can be shown to officially 'start', the less blame can be put on politicians, scientists and civil servants. But government inquiries into food and farming disasters are not really about apportioning blame at all.

The Phillips Inquiry

What Richardson had uncovered, however, was the beginning of the BSE crisis that would help to topple the Conservative government. It would alter people's perceptions of food safety forever. As the stench of cover-up, collusion and incompetence began to engulf the Tories, many people – including affected families – called for a public inquiry.

To their credit – and with little political risk to them – the new Labour administration agreed. The man asked by Labour to uncover how it had all gone so badly wrong was Lord Phillips of Worth Matravers, a High Court judge. Phillips was to chair one of the most contentious public inquiries in post-war Britain. The inquiry took evidence from over 900 witnesses, including scientists, former ministers, senior civil servants and families of victims, and reviewed over 5 million pieces of paper that were condensed into a 16-volume report. It cost £27 million, was meant to last a year, but took two years and nine months.[59]

It was inevitable, perhaps, that an inquiry into such a controversial subject would be controversial itself. The inquiry was meant to finally uncover the truth of what happened with the BSE saga. There is no doubt that it brought together a vast volume of information on BSE that has been useful in understanding the disease, but it has been widely criticized. First, many of the victims' relatives wanted ministers or civil servants blamed, but they were not. 'Any who have come to our report hoping to find villains or scapegoats, should go away disappointed' argued Phillips.[60] Secondly, it was criticized for not understanding the fundamentals of TSE science and thirdly how it processed and presented the information to the public.

'Lord Phillips deliberately excluded from his team anyone with direct involvement in TSE research', argues Dr Alan Dickinson, resulting in the science section being 'unsatisfactory'. The inquiry would later criticize several influential BSE committees for not having an expert on them, and yet the inquiry did not have its own expert.[61] Dr Millstone adds 'the Phillips report is a classic official British document in that the executive summary at the front bears only a tangential relationship with the evidence that it purports to summarise'. The report, when it was finally published in 2000, was seen as being full of contradictions. Three of the key conclusions are that:

1 'the government did not lie to the public about BSE;
2 most of those responsible for responding to the challenge posed by BSE emerge with credit;
3 in dealing with BSE, it was not MAFF's policy to lean in favour of the agricultural producers to the detriment of the consumer'.[62]

But it is difficult to see how these three conclusions marry with the facts and hard evidence. The government ignored the evidence from independent scientists, who were vilified. Nearly every expert working within the field of TSEs in the late 1980s believed that it was possible that BSE could be transferred to humans, it was just not yet proven. We now know that the government did not commission crucial research that would have told them whether humans were at risk. Economic interests overruled consumer concerns.

Even Phillips' report shows how the government response was typified by unnecessary time delays, and how the public was not informed of the changing perceptions of the risk of BSE being passed to humans. Phillips laments the government 'ignorance' on BSE and the 'serious failure to give rigorous consideration to the amount of infective material' going into the food chain

Phillips failed to reconcile the feeling of public 'betrayal' felt about Dorrell's historical announcement with the perception that major players 'emerge with credit'.[63] The inquiry was seen as being so lenient on the major players that the report was dubbed a whitewash.[64] As the *Sunday Times* said when the inquiry was published, the 'lack of criticism is sometimes baffling'.[65] The report was at odds with public perception at the time. A poll for the Consumers' Association in April 1996 found that 70 per cent of respondents thought the government had withheld information regarding the risks of BSE and two-thirds believed that food producers had an influence over government policy that was against consumer interest.[66] Essentially two-thirds of consumers felt that economic interests were being given priority over protection of the consumer, yet Phillips quite inexplicably denies this. The evidence seems otherwise, though.

MAFF – the 'Ministry of Truth'

The ministry with primary responsibility for handling the BSE crisis was MAFF, a ministry with a reputation for looking after its own. In the 1980s and 1990s MAFF was a powerful government department, as its name suggests, controlling agriculture, fisheries and food. It has also been said by its critics that its name reflected its order of priorities – first agriculture, second fisheries and lastly food.[67]

Until it was abolished after the foot and mouth crisis, MAFF was supposed to act on behalf of both farmer and consumer. This dual role was often criticized during the BSE crisis. It was obvious who won; it was a ministry that looked after the agricultural trade.[68] Edwina Currie, the former Tory Health Minister who resigned over salmonella in eggs, said that MAFF officials were not 'the least interested in public health and felt that their task was to look after the farming industry'.[69]

Critics of the government argue that MAFF's stance was flawed. Dr Iain McGill, a vet who quit the government after being warned not to tell anyone about what his research had uncovered, summed up MAFF's approach to BSE by maintaining that 'absence of evidence' is not the same as 'evidence of absence'.[70]

Professor Philip James, who sat on the EU BSE Scientific steering committee, believes that MAFF's response was typical of other agricultural ministries around the world. 'Their primary objective was to sustain and nurture these poor farmers whose livelihood was desperate and on whom the nation depended. We forget that this was their whole concept.'

Time and again economic interests would steam-roll consumer concerns. The Whitehall machine went into overdrive as ministerial reputations and that of the beef industry superseded worries over public health. The industry came first, the concern for exports overrode everything else. In one memo, a MAFF official noted how: 'If the disease turned out to be bovine scrapie it would have severe repercussions to the export trade and possibly also for humans'. Rather than be open, a cloak of secrecy engulfed MAFF. By 1987 there was a 'total suppression of information on the subject', the principal reason being 'the possible effect on exports and the political implications.'[71]

However, MAFF did not look to find the answers and did not fund key scientists who could have provided tests that would have been able to detect BSE in the herd (see Chapter 3). 'I don't think they were interested in anything at all but politics,' says Professor Lacey now. Lacey, who, in Orwellian speak, has dubbed MAFF the Ministry of Truth, believes that once BSE became public, 'MAFF's priority was to do enough cosmetic research to show that it was being done. They couldn't do very much, otherwise they would have to admit there was a big problem'.

In fact one of the greatest flaws of MAFF's response to the BSE crisis was insisting that BSE was scrapie in cattle, but the ministry never set up what would have been a simple experiment, which would have been to feed scrapie-infected sheep to cows.[72] The line was simple: if BSE was scrapie in cattle, then it was unlikely that it was going to be transmitted to humans because scrapie had been endemic in the UK flock for 200 years without ever infecting humans. So without any evidence to the contrary, the ministry could carry on denying any risk to humans.

There are examples of MAFF stopping the publication of research at critical times. In September 1987, when a MAFF researcher suggested writing about the new disease in the medical press – either the *British Medical Journal* or *Lancet* – he was told, 'Not at present. It would over emphasize the possible link to human spongiform encephalopathies'. The following month MAFF published the results of the first known case of BSE in the veterinary press instead. It would take an independent doctor to first suggest a link between BSE and vCJD in the medical press in June 1988, a crucial eight months further down the line.[73]

Leading TSE scientists at the Neuropathogenesis Unit (NPU) in Edinburgh who did not work for MAFF were also threatened that if they did say something publicly they would lose their jobs. So not only was the key research not done, or was not published, but the research which was done was undertaken for a reason. Suzi Leather, who was on MAFF's Consumer Panel, distinctly remembers 'at one panel meeting the Chief Food Scientist, Dr Howard Denner, saying that on the whole research was commissioned in order to protect ministers'.[74]

As MAFF made the primary decisions, lurking in the background was the Treasury, keen to keep the purse strings tight. 'Its dark presence lurks throughout the whole BSE saga as insidiously as the agent of the disease itself', says Hugh Pennington, Professor of Bacteriology at the University of Aberdeen, who chaired an inquiry for the Scottish Office into food poisoning in Lanarkshire in 1996–7.[75]

'The real culprit is the British Treasury', adds Hugh Fraser, from the NPU in Edinburgh.[76] MAFF and the Treasury's underlying effort was to 'maintain the status quo of beef and dairy consumption and beef exports,' argues Professor Lacey. 'That has been their priority all along and it still is. Although now of course, half the animals are being destroyed and the export market has gone. But they are still trying to maintain the status quo.' Maintaining the status quo meant suppression of information that the public had a right to hear. It meant denying the evidence as the young started dying. The grandmother of the first teenager to contract vCJD, Vicky Rimmer, was told simply why she could not talk to the press: 'think about the economy, think about the EEC'. She could not believe what she was hearing. 'My reply was "For God's sake, this is a child's life."'[77]

Vicky had been lying in a coma for 20 months when Stephen Dorrell made his announcement to the House of Commons in March 1996. Two months before Dorrell's announcement, in February 1996, another tragically young vCJD victim, Peter Hall, had died just days before his twenty-first birthday. Peter had been a vegetarian for the last five years of his life, as he could not stand cruelty to animals. He was bright, fit and studying environmental science at university. Within months he was a staggering, incoherent wreck whose life was just ebbing away. The day before Dorrell's announcement Peter Hall's family was told he had died from vCJD. An inquest would later rule that Peter's death was 'more likely than not, ... caused by eating BSE infected beef products'.[78] Vicky Rimmer died in November 1997. 'I believe that someone in the Government at the time knew what was going on but turned a blind eye', Vicky's grandmother told the Phillips inquiry. 'I now feel that someone should be answerable and should take the blame and be made to answer for those persons who have died as a result of new variant CJD'.[79]

But Peter's and Vicky's families still wait for justice. The BSE saga is one that has been typified by cover-up and secrecy rather than openness, by vested interest rather than public interest and by arrogance that defies belief. After Dorrell's announcement, Arthur Beyless, whose daughter died of vCJD, met Dorrell, who happened to be his local MP. 'He never said a dicky bird about being sorry', says Beyless, 'and he never will'.[80]

When Richard Packer, the Permanent Secretary of MAFF, gave evidence in 1996 to an EU inquiry on BSE he was asked: 'Was the British Government responsible for the whole disastrous mess?' 'No it wasn't,' said Packer. 'In the main it was an act of God, if you like. Some of the things we have done we might with the benefit of hindsight have done differently. To that extent the British Government accepts responsibility – but that is not very much of the responsibility'.[81]

'These people have no remorse' responds Dr Harash Narang. Politicians and civil servants, he says, 'were given the impression that the incubation period will be 20 to 40 years and therefore they would not see the problem in their life-span. This is what they thought. Unfortunately it happened right in front of their eyes'.

Cover-up and censorship?

So was there a deliberate cover-up? Critics point to the omission of two key organizations that had expertise in TSEs and infectious diseases as evidence that there was a direct cover-up at the heart of Whitehall. The organization that handled the initial BSE carcasses was the CVL of the State Veterinary Service (SVS). This was a MAFF department, answerable at the time to the Chief Veterinary Officer. MAFF was in control of the crisis.

When the outbreak broke, control would be given not to the NPU in Edinburgh, which had the leading experts in scrapie and TSEs in the UK, but were not under the control of MAFF, but to the CVL, which were under MAFF control. Alan Dickinson believes that if the experts at the NPU had been given control, then the size of the epidemic 'would have been considerably smaller.'[82]

But it was not only the NPU that was barred from involvement. Professor Lacey now says that: 'I am now absolutely convinced that from 1988 for the next two years, there was a deliberate conspiracy. It was orchestrated. The problem then was, by the time they realised the scale of the problem they couldn't afford to do anything about it. That is still the case today. The most significant thing at all was the failure to involve the Public Health Laboratory Service'. Lacey adds that: 'The Public Health Laboratory Service [PHLS] is answerable to the Department of Health and quite clearly when there is infection in a food animal, there is a prima facie case of a potential hazard to the human population. They did not allow any involvement of the PHLS'. The PHLS did not get involved in the BSE saga until after Dorrell's announcement in March 1996.[83]

Statements from senior PHLS officials show that Government Ministers prevented PHLS involvement in the BSE crisis. 'It was made increasingly clear to me that DoH [the Department of Health] and Ministers did not wish the PHLS to work upon BSE/CJD, nor to be seen to work or comment upon the subject', wrote the Director of the PHLS from 1985 until 1992, Sir Joseph Smith. 'This caused me much concern', said Sir Joseph, who believed the PHLS should be involved in the critically necessary human epidemiological studies of BSE/CJD. Smith raised his 'concerns' with DoH officials, but received a crucial letter from Dorrell in 1991 that 'addressed the question of spongiform encephalopathies'. It was this letter that the PHLS understood to be confirmation 'that we should not be involved in this work'.[84]

The current Director of the PHLS, Dr Diana Walford, was also told that from the time she was appointed in 1993 to 20 March 1996, 'there was no requirement for PHLS involvement in spongiform encephalopathies'.[85] This meant that the organization whose remit is to protect the population from infection was not working on a lethal transmissible disease. This was because those in power dismissed the 'solid evidence that humans could be susceptible to TSEs by eating contaminated food,' argues Dr Will Patterson, a Consultant in Public Health Medicine, from the North Yorkshire Health Authority. It is Patterson's opinion that 'the absence of BSE from the corporate agenda of the PHLS ... seriously undermined the effectiveness of the United Kingdom public health framework to protect the population from infection'.[86]

A slow policy of sedation

The government was so obsessed with economic considerations, that on the whole their response was excessively slow. The first official case of a cow suffering from a 'novel progressive spongiform encephalopathy' recorded in the *Veterinary Record* is the one on which Carol Richardson performed the autopsy and the date is given as April 1985, although Richardson did not identify a 'spongiform' disease until September 1985. So if we take April 1985 as our starting point when BSE was first officially realized by vets, it would not be until December 1986 – some 20 months later that scientists at the CVL confirmed this. They then notified the Chief Veterinary Officer of the new disease that month. From this moment, officials and Ministers followed an approach whose 'object was sedation', concluded the Phillips Inquiry.[87]

There was a further six months delay – until June 1987 – before scientists at the NPU in Edinburgh were formally consulted over the disease. There was then a further delay until 31 October 1987 when the Government admitted that BSE existed, with experimental results published in the *Veterinary Record*.[88]

By January 1998 with the number of BSE cases rising, MAFF had to act and considered making BSE a notifiable disease as well as introducing a slaughter and compensation scheme. 'We do not know where this disease came from, we do not know how it is spread and we do not know whether it can be passed to humans,' one MAFF official noted privately. 'We have to face up to the possibility that the disease could cross another species gap.'[89]

However, MAFF was preoccupied with getting the industry to pay for any slaughtering programme,[90] as well as with continuing worries over exports,[91] so six more months went by before a compensation scheme was introduced. When it was, in August 1988, a farmer was only paid 50 per cent of an animal's value.[92] By only offering 50 per cent rather than 100 per cent MAFF sowed the seeds for the BSE crisis turning into a disaster.[93] It meant that many cows went unreported – for when the full price of a cow was paid in February 1990, the number of cows with BSE jumped by 73 per cent.[94]

It was only in March 1988, just under three years from the first officially recognized case, that MAFF wrote to the Chief Medical Officer seeking advice on the risk of BSE to human health, a delay seen as 'serious' by DoH officials. Internal memos written at the time show there was preoccupation with civil servants to 'play down the human health issue'. Finally that month, MAFF set up an Expert Committee to examine the crisis.[95]

The Southwood Committee

When MAFF set up their first independent advisory group in April 1988, it was led by Sir Richard Southwood, who was a Professor of Zoology at Oxford University. Southwood started a key list of scientists from the Departments of Zoology at both Cambridge and Oxford, who had been asked by the government to play instrumental roles in the BSE, GM and foot and mouth disease debates, but whose expertise is not primarily in food or farming.

Many of the 'false reassurances' which plagued BSE during the early years stemmed from the Southwood Committee, whose Working Party were 'deliberately chosen' as people who were not TSE specialists.[96] The Committee has been called a 'a dire episode in a terrible saga'.[97] One natural choice would have been Dr Alan Dickinson, the leading expert on scrapie. But he was not invited to join or asked to give evidence to the Committee.

As Professor Hugh Pennington explains, that Dickinson 'was excluded was not only bad science (science draws its strength not from making hypotheses but from testing them), bad administration and bad politics, but it was bad for public health. It almost certainly meant that people went on being exposed to the BSE agent for longer than necessary.'[98] Basically, Dickinson's absence cost lives.

When Southwood's report was published it was both contradictory and reassuring at the same time. He said the risk of transmission of BSE to humans was 'remote'. 'From the present evidence, it is likely that cattle will prove to be a "dead-end host" for the disease agent and most unlikely that BSE will have any implications for human health.'[99] It is the 'dead-end' statement that many of the government's critics, like Professor Lacey, find so amazing. 'The name TSE means transmissible', he says. 'It was so incredible to say from the evidence that cattle were a dead-end host, when there hadn't been enough time for the other evidence to accumulate. And they knew it was transmissible by experiments in mice. They quoted that. It is unbelievable.'

The words 'dead-end host' and 'remote risk' would be used to reassure consumers for years that beef was safe. Continuing until Stephen Dorrell made his announcement in the House of Commons in 1996, it was 'cited as if it demonstrated as a matter of scientific certainty, rather than provisional opinion, that any risk to humans from BSE was remote'.[100] As beef was pronounced safe, lost in the small print was a sentence which stated that if the committee were 'incorrect, the implications could be extremely serious'.[101]

There were inconsistencies too. Southwood's working party had been informed that that the placenta of infected cattle were particularly infectious.[102] Yet the Southwood report 'assumed' that there was no vertical transmission, and for years afterwards the government itself refused to admit that 'vertical transmission' might be an issue. It was typical 'doublespeak' says Lacey. 'Cows could not pass the disease down to their offspring – yet their placentas, and therefore inevitably their calves, might be infected.'[103] At the same time the Committee said, as a precautionary measure, that 'manufacturers of baby foods should avoid the use of ruminant offal and thymus; the latter can currently be described on food labels as meat'.[104] The simple question was, if babies were susceptible, why not adults and what was the safe age to suddenly eat beef?

There was infighting between the DoH and MAFF as to the report's wording. MAFF asked for key sections of the draft to be re-worded. Sir Richard even admitted later that he had been pressurized by MAFF not to recommend anything 'which would increase public expenditure.'[105]

'What they should have said' says Lacey, is that 'we have a new disease. It is infectious. We know that these infections cause disease in a range of mammals. We don't know what range of mammals BSE can infect. Therefore we should assume the precautionary principle, that it may affect us unless proved otherwise. Because we human beings are vulnerable'.

Neither Dickinson's nor Lacey's views were sought. They would have been two of the key scientists who could have changed the course of the BSE epidemic. But the government had handpicked a team of scientists with no direct knowledge of TSEs. Their report was based on flawed assumptions and recommendations were largely 'guesswork', Sir Richard conceded later. Yet it would become the Government's 'bible' on BSE.[106] The report justified the policy of inaction whilst the epidemic grew.

Short-term economics continue to overrule long-term health

What action the government took was often flawed. Three months after the setting up of the Southwood Committee, in June 1988, the government finally made BSE a 'notifiable disease' and introduced the Ruminant Feed Ban, whereby they stopped ruminants being fed to ruminants, that is cows. But it still allowed the continued use of sheep and cattle to be rendered into feed for pigs and poultry, even though no one knew whether they were susceptible to BSE. It would have made more sense to ban it outright, but then the renderers would have been deprived of their principal markets.

The decision had devastating consequences. For approximately the next six years cross-contamination continued between feed destined for cattle and feed destined for other animals, greatly prolonging the BSE epidemic.[107] Over 41,000 cattle that have since developed clinical signs of BSE were infected after the ruminant feed ban came into effect.[108] It also allowed pigs and poultry to spread the disease on farms, despite the fact that MAFF knew about Dickinson's research that had shown that healthy animals could excrete infection.

By 1989, it was announced that the Specified Bovine Offal (SBO) order would be introduced that would ban what was deemed to be the most infectious tissues – brain, spinal cord, spleen, thymus, intestines and tonsils – from going into the human food chain. But no one really knew what the infectious agents were because the research had not been done in cattle.[109] Research on other species had shown that TSEs infected a host of other tissues such as nerves, eyes, lymph nodes, kidney, lung, uterus, blood and muscle.[110] Critics argue that the tissues selected for the ban were chosen because they had the lowest commercial value and they were easily removed.[111]

It was not until September 1990 that the SBO ban was changed to include the use of any specified bovine offal to all animal feeds. However, there was no system put in place to check the slaughterhouses and renderers to see if the rules were being obeyed, so the Animal SBO ban was effectively unenforceable and was widely disregarded for years.[112]

But still MAFF's policy was flawed. Once all the meat had been removed from the carcass, an industrial pressure cooker was used to recover any remaining meat. It was made into a slurry called Mechanically Recovered Meat (MRM) which was classified as real meat. The major source of bovine MRM was the spinal column, which is highly infectious.[113] Recent research shows that 5000 tonnes per year went primarily into burgers and frozen mince, much of it used for school dinners, a fact labelled 'repulsive' by Lester Firkins from the Human BSE Foundation.[114] It was not until 1995 that action was taken to stop MRM going into our diets.[115]

But it was not until after Dorrell's 1996 announcement that there was a ban on all mammalian protein to all farmed animals. It had taken ten years for the government to remove pathways of infection it could have removed from day one. But infection pathways remain, as do contamination pathways. It is still legal, for example, for the pet food industry to import, from the continent, MBM that could be contaminated.[116]

Scientists do not know if it is the continuing use of contaminated feed or vertical transfer or even contaminated pasture that means that BSE's 'third phase' continues. 'My hunch is that in 10 years time we're going to have scientists and politicians neurotically wondering about at what level do you accept the occasional BSE animal entering the food chain', says Professor James.

What really scares the scientists is that they do not know when it will end. 'The epidemic is not over,' SEAC's Professor Smith told the OTM meeting. 'It is difficult to predict when the epidemic will be over.'[117]

CHAPTER 3

Treated with Derision

'The official line that the risk of transmissibility was remote and that beef was safe did not recognise the possible validity of any other view. Dissident scientists tended to be treated with derision …'

The BSE Inquiry[1]

'They thought that by locking me out – by padlocking me out – the problem would go away. But it hasn't'

Dr Harash Narang

The BSE saga is littered with scientists who were marginalized or attacked for speaking out. Although some – but not all – of these scientists were criticized for being alarmist, on the whole they have been proved chillingly correct in their analysis. If their voices had been heard and acted upon, we know that the BSE epidemic would have been very different. It might not have happened at all.

Dr Alan Dickinson

The most influential committee of the BSE crisis was the Southwood Committee, 1988–1989, because it allowed politicians to argue for years that the risk of transmission of BSE to man was remote. The committee, in making that decision, had 'drawn comfort' from the belief that BSE was probably scrapie in cattle and therefore posed no risk to humans as scrapie had been around for over 200 years and was thought never to have infected humans.[2]

This false reassurance could have been prevented if the committee had included Dr Alan Dickinson, a leading expert on scrapie, who had been working on the disease since the mid-1950s in Edinburgh. His work remains pivotal to the understanding of TSEs. Dickinson started to search for genetic controls of scrapie in a wide range of mice and to search for any strains of the agent. He showed that scrapie was not caused by a sheep gene, as was the contention at the time and then went on to look for different strains of scrapie, leading to different incubation and infection rates. He eventually found over 20 different strains.[3]

Dickinson's analysis would have been invaluable to the government. But he was sidelined. It is conceivable that the government do not like Dr Dickinson because he is a principled scientist, known for fighting the Whitehall elite with his blunt talking and precision analysis. He has criticized the way governments have run roughshod over science, notably over the increasing politicization of research councils and cutbacks to research budgets.

Dickinson was the brainchild behind and subsequently first Director of the Neuropathogenesis Unit (NPU) in Edinburgh. Started in 1981, it was jointly funded by the then Agricultural and Food Research Council and the Medical Research Council. Dickinson believed that independence was necessary to undertake cutting edge scientific work. On being told that the Unit was to be merged into the Institute for Research on Animal Diseases, which later became the Institute for Animal Health, and would be coordinated from hundreds of miles away in Berkshire, Dickinson resigned.[4]

'As there seemed to be no rational end to the research shambles that had been imposed, I chose to retire 2 years early at that stage, weary and sad for my NPU colleagues, but without regrets', he wrote.[5] He was sidelined and so were his colleagues, although other members of Dickinson's team in Edinburgh, Moira Bruce and Hugh Fraser, both world experts, have developed the best strain typing tests and produced the most convincing results linking BSE and vCJD.

Dickinson had been proven right over TSEs before. In the mid-1970s he had realized that the practice of extracting the growth hormone from pituitaries taken during autopsies and using them to treat dwarfism could be creating a risk of CJD. 'I predicted there was a real risk of growth hormone being contaminated with CJD, seven and a half years before the first case world-wide', he says.

So what could Dickinson have told the government? First, his knowledge alone, as the leading scrapie expert in his field, would have been invaluable. He had many of the pieces of the jigsaw already in place. Dickinson had worked on the species transfer of TSEs between sheep, goats and rodents with attempted transfers to mink and primates. If he had been granted permission he would have looked at the transfer of the disease to cows too, but he was stopped.[6]

Secondly he would have told them that the reassurance that Southwood took about scrapie not being transmitted to humans was wrong, as transmission was possible, although scrapie – or a TSE coming from sheep – is not very easily transmissible to humans. 'I think the least likely option is that BSE came from sheep', says Dickinson. 'There is not a shred of evidence that scrapie does not transmit to humans, there is very good evidence that it is never transmitted under quite reasonable exposure times, it is never transmitted under such a focal pattern, that it has drawn attention to itself'. Key experiments looking at whether a species barrier existed between sheep and cows were never set up, but would have shed light on whether it was only infected cattle tissue that had caused the epidemic.[7]

Thirdly, decades of research had taught Dickinson that there are no easy answers with TSEs. Even the word is a misnomer. 'The word 'spongiform' puts a profoundly wrong understanding on the disease' he argues, because in some strains of TSEs you don't get the characteristic holes at all. An important but 'misunderstood issue is that I have known of medical neuropathologists from twenty years back say that it could not have been CJD because there was no vaculation, it could not have been BSE because there was no vaculation'. This begs the awkward question of how many samples have been misdiagnosed because the typical vaculations were not there.

Fourthly, for years MAFF maintained that vertical transmission was not possible with BSE. This false hope undermined the government response and continues to prolong the BSE epidemic. Dickinson's team had concluded otherwise; that 'maternal transmission of the agent could play an important role in the incidence of the disease'.[8] The government also ignored his work on pastoral contamination outlined in Chapter 2.

Fifthly, Dickinson could have influenced the direction of research. 'There is so much that has gone wrong,' he says. The Committees set up to look at BSE made 'disastrous decisions', according to Dickinson. 'I think it is fair to say that considerably more that half of the £60 or £70 millions put into BSE research was forseeably wasted, and I think foolishly', he argues, although the situation has improved recently. 'This is a view shared by many of those with proven TSE-research expertise, some of whom are still involved in the research and therefore too prudent or intimidated to publicize their opinions'.[9]

Much of this research has been spent trying to understand the basics of the prion protein, known as PrP. But Dickinson is one of a number of leading scientists who believe that the 'prion' theory – that the agent is an infectious protein – 'is unable to account for a growing body of evidence'.[10]

As the research goes wrong, the answers lie unresolved. One unanswered question is the actual route of transmission. 'I have no doubt that infectivity goes across the dorsal gut in cattle', says Dickinson. 'But do you know that that is the route of infection that infected those cattle in a lethal sense?' He says we still do not know whether the infection route for cows with BSE was the dorsal gut or another route. 'If you find one route that is a possibility you cannot assume that is the lethal route. You see how little we know.'

There is an on-going debate that infection through cuts and grazes may be a much more efficient infection route than actual ingestion. Dr Dickinson points out that other senior TSE scientists agree with him. Carleton Gajdusek, who received a Nobel Prize for his work on 'kuru' among the Fore tribe in Papua New Guinea, conceded that villagers could have been infected by the sharp bamboo edges that make up their cooking pots rather than through cannibalism.

Dickinson also says there is evidence from mink. The pioneering researcher on mink was Dick Marsh in the USA. In one case in the mid-1980s, some 60 per cent of mink had died after eating infected cattle, a very high infection rate. Marsh told Dickinson that he thought he knew how they were getting infected. Their food was placed on the top of wire cages and the animals would lacerate their mouths as they tried to get to the food.

When Dickinson experimented by scarifying a mouse with infected tissue, he found that it was almost as efficient a route of infection as injection straight into the brain. He believes teething may be an important infection route. 'The UK is different from all other countries in presenting far younger calves with slurries based on MBM'. If you get bleeding with tooth eruption, with damage to the oral cavity, he says, 'that may be the highly efficient route for the UK outbreak. Equally you may say that here you have vCJD in young people and one thinks that tooth eruption and things like that could be an explanation'.

It is this issue that makes Dickinson believe that there could be more vCJD cases than are predicted, although he stresses the need for caution. Dickinson believes that above all else research needs to look at natural routes of transmission and relative infectivity. 'The unresolved question is whether the TSE strain that causes BSE and vCJD is intrinsically more easily transmitted to other species, including humans, than other TSE strains' he says.[11] To date that question remains unanswered.

The British government would do well to seek Dickinson's advice on another issue of concern, whether the British sheep flock is harbouring BSE or just scrapie and the merits of the proposed Scrapie Eradication Plan. Dickinson believes that the attempt to identify a gene that would 'resist' all known (and future) strains of TSEs, may not be realistic, and that there are risks 'of over-narrowing the genetic diversity of the national sheep flock'.[12] What happens, in the future, asks Dickinson, 'when a strain is encountered that is able to infect these supposedly resistant sheep? At that stage, will all the UK sheep be open to infection, as was the case when the entire cattle population encountered the BSE strain?' Only time will tell.

Dr Tim Holt

At the same time as Dickinson was excluded from the Southwood Committee, a young doctor, Tim Holt, who had just graduated from St George's Medical School in London was become increasingly concerned about the link between BSE and vCJD. Looking back he says he was 'young and naïve', but felt he had no option other than to speak out as there was an 'enormous risk of a health problem' and 'something needed to be said'.

'I did wonder why other people were not more concerned about it,' says Dr Holt, who is now a rural GP in Yorkshire. 'There was a lot of scientific uncertainty, but the politics of it was that anyone who suggested that it might be unsafe was threatening to undermine the entire beef industry and everything that was attached to it.'

Along with a dietician Julie Phillips, Holt published an article in the *British Medical Journal* (BMJ) in June 1988 that suggested that BSE might prove a risk to humans. This was the first time that anyone had voiced such concerns in the scientific literature. Government scientists knew about the risks and their silence was deafening. Holt and Phillips pointed out that 'there is no way of telling which cattle are infected until features develop, and if transmission has already occurred to man it might be years before affected individuals succumb'. They concluded that 'it is possible, but unproved, that many asymptomatic cattle are nevertheless as infective as those symptomatic animals which are immediately destroyed for public health reasons. So should not the use of brains in British foods be either abolished outright or more clearly defined? Then in the absence of more compelling evidence those of us who wish to exclude it from our diets at least have that choice.'[13]

'I knew that a lot of the offal got into our food', says Dr Holt now, 'but we still had a difficulty persuading people that this actually happened, they still thought that when they had a meat pie that all that was in it was lean beef. They didn't know about mechanically recovered meat.' Mechanically recovered meat or MRM was a continued source of potential infection until the mid-1990s.

Holt expanded on his concerns to the BSE Inquiry. 'To reassure me that there was no risk to human from eating offal, in particular the brain and central nervous system tissue from animals with this new disease, would have required information which frankly could not have been possibly available at the time.'[14]

Holt, too, was vilified. 'During 1988 when the paper was published it was still an off-the-wall, slightly dangerous thing to be saying,' says Holt, recalling how he was called alarmist and attacked by both MAFF and the meat industry. One MAFF scientist said that Holt's criticisms of MAFF's delay in responding to the first case of BSE was 'inappropriate and reproachable'.[15] However, some scientists were interested in what Dr Holt was saying – for example, Hugh Fraser, one of Dickinson's colleagues from the NPU. 'Our work on scrapie highlights our lack of knowledge such as the levels of infection of healthy cattle incubating the disease', he wrote to Holt. 'The incorporation of lymphoreticular tissue, such as spleen into meat products is just one example of a policy which may be difficult to explain.'

Fraser also informed Dr Holt that he had transferred BSE to mice, another example of the disease being able to cross the species barrier. Holt visited Fraser in Edinburgh in November 1988, during which time Fraser let slip that people at the NPU had changed their diets to avoid offal from beef.[16] 'Why', thought Holt, 'if the scientists had stopped eating meat products, were we eating it in our foods bought from supermarkets?'

Dr Hugh Fraser wrote to Holt the following month confirming something the politicians were to deny for a further eight years: that transmission from animal to human was possible. 'It is plausible', he wrote, 'that a slow virus encephalopathy distinct from CJD, could arise in humans from dietary exposure to an animal virus'.[17]

In a letter written to the Medical Research Council after his return from Edinburgh, Holt expressed his growing concerns. By now Holt believed the public was 'wide open to a similar epidemic [to BSE], given the known oral transmissibility (in the lab) of the scrapie agent, the laws governing the use of offal in human food, and the current Government policy on compensation which is set in such a way to encourage concealment of the disease by farmers'.[18]

'I didn't even get a reply', he says. 'I thought at the time, that surely, surely, the Medical Research Council should be more concerned about this.' But they were not and neither were MAFF. Holt later gave evidence to the Phillips Inquiry that MAFF 'was less interested in investigating the problem than they ought to have been'.[19]

Holt also expressed his 'concern ... that in an effort by the Government to protect the British meat industry, the potential medical importance of the issue may be ignored or forgotten'.[20]

When Holt wrote the BMJ piece, the government had not yet set up the Southwood Committee. When it was established Holt was 'amazed' that none of the Edinburgh NPU scientists – like Fraser or Dickinson were on it. 'The other person who was missing was Kenton Morgan, who is now a Professor at Liverpool [of Epidemiology in the Faculty of Veterinary Science], but was a research vet interested in scrapie' he says. Morgan was an obvious candidate for the Southwood Committee, argues Holt: 'Why wasn't he selected, because they know what they would get if they asked him for their advice. It would have been the same sort of line they would have got from me, but it would have been more difficult to ignore him. There was a lot of bias in selection of people for these committees. I don't mean that as criticism of those who were selected. It is more about the omissions.'

Professor Morgan was another scrapie scientist marginalized by the government. In his evidence to the BSE Inquiry he said that: 'It was also clear to me at that time [1988] that the major question, in considering the risk to people, was not whether there would be transmission but how many cases there would be.'[21] So, effectively, Morgan warned that it was only a matter of time before BSE would be transferred to humans, eight years before the government admitted it. In response, Morgan was blocked in his research efforts and became 'unsuccessful in obtaining grant funding for TSE work in spite of considerable success in other areas of veterinary epidemiology at that time'.[22]

When Southwood's report was published, Holt appeared on the BBC's Radio 4 *Farming Today* Programme. Faxed the report the night before, he thought that the most 'suspect conclusion' was the 18-month old age limit under which certain foods should not be fed to babies. 'I was quick to point out that there was no basis for that particular age group as being particularly at risk.' Holt was also troubled that Southwood was not concerned with offals from pre-clinical cows. So a 'mere' doctor appears to have had a greater understanding of the problems than those on the committee.

Holt continued to follow BSE closely and felt 'sickened' to hear MAFF's 'over confident' line 'that they had the whole thing under control', at a conference in 1993. 'I thought that by 1993, with all that had happened they would be a little bit more humble. I realised that until we actually had evidence that BSE was transmitted to man, it was not likely to change.'[23]

Holt says he 'expected more support from the Phillips Inquiry' too. He believes that the overriding conclusion to be drawn from Phillips is that, whilst people like himself turned out to be correct, they were lucky because the evidence was not particularly worrying. 'That to me is wrong. I was there. I remember what Hugh Fraser felt and what Kenton Morgan felt. It is wrong to suggest that there was not enough evidence at the time to be concerned. There was. It is that just enough reassurance was extracted from these people on these committees, including Southwood, so politicians can now claim that they have fulfilled their responsibility to the public.'

On the whole the Phillips Inquiry 'gave a lot of people a very easy time', says Holt. 'It is very regrettable' he feels, 'that all those years went by during which BSE was common, but it was not being effectively removed from human foods.'

Dr Helen Grant

One person who had been alarmed by Holt's letter to the BMJ was Dr Helen Grant. Another key critic of the government, Grant was a Consultant Neuropathologist at the Middlesex and then Charing Cross Hospital. Grant is now an octogenarian, with a crystal clear recollection of the BSE saga.

Through her interest in slow viruses (particularly with reference to multiple sclerosis), Grant was one of only a few people to be aware of both scrapie and CJD in 1988. 'I knew all about CJD as every neuropathologist does,' she recalls now, 'but I also knew about scrapie'.

She told the Phillips Inquiry that she was 'horrified because I suddenly realised why Government assurances about BSE – namely "we have lived with scrapie for two-and-a-half centuries and it has not done us any harm so we won't have any trouble with BSE" – were based on a false premise'. This was 'that cattle brains and sheep's brains were dealt with in the same way in the abattoirs which they obviously were not'. Grant believes that ' humans have never been seriously exposed over the centuries to the scrapie agent,' as sheep brain's are seldom removed to be eaten. 'Simple economics is the reason: sheep's brains are too small to make the intricate process of their removal worthwhile.'[24]

Grant criticizes the government for 'burying information' and 'doing nothing for eighteen months,' after the first cases of BSE. Finally, she says, when the Southwood report was published, 'MAFF stopped feeding cattle the infected brain material, but it did not cross their minds that we might be in the same boat, we humans – brilliant isn't it. Then they didn't do anything about our food. Almost everything they did was wrong'.

Like other government critics, Dr Grant was alarmed that the Southwood Committee said the risk to humans from BSE was 'remote'. The day the Southwood Report was published, she said: 'Of course humans are at risk of catching this disease, because experimental work on scrapie infected brains has revealed that it is very easy to infect almost any mammal including apes and we are just another ape. We must not be made to eat cattle brains, which at the moment go into our foods'. But she says now, 'they paid no attention of course. They did nothing about it.'

In March 1989, Grant wrote that the Government was concentrating on baby food 'to divert the public from thinking about other foods and thus to imply that they are safe, which they are not'. The official Inquiry credits Grant's article as one of the influences combined 'to drive MAFF towards the decision to introduce a ban on using for human food those types of offal that were most likely to carry BSE infectivity'.[25]

Grant maintains that she spent the rest of that year – 1989 – 'hammering away' at the Secretary of State for Agriculture, who was at the time John MacGregor. She was told 'not to worry', receiving 'only short and reassuring replies containing what I believed to be inaccurate information'. Officialdom, she says, treated her 'early warnings with hostility'.[26] She was to be dismissed later as being 'too out of date'.[27]

Grant also became 'increasingly aware after 1988 that questioning official dogma about BSE brought difficulties to one's career. I was myself about to retire from the Charing Cross Hospital, where I worked as a Consultant Neuropathologist, but I observed with horror that the good reputations of dissenting scientists in the field, not least Dr Stephen Dealler and especially Dr Harash Narang were systematically undermined'.[28]

Like Lacey and Dickinson, Grant also believes that teething may have created a more lethal transmission route than eating. 'One of the features of this disease is that it has attacked young people, whereas normal CJD attacks older people. The reason for that is that when we were getting infected food from 1981 until 1989 many that have now died, were children,' says Grant. 'Therefore they were shedding teeth and therefore there were raw areas in their gums, where the teeth go. It means that mouthfuls of infected meat pies could go straight into the blood stream. Therefore the incubation period is very short, and when they blossomed out into full disease they were still very young.'

'There are others who got it the routine way, through the stomach, which takes very much longer. When they blossom out into the disease they are going to be much older. I don't think it is going to explode, but the death toll is going to continue rising. Nobody knows of course, but we were all exposed to it during that era.'

Also, like the others, Grant is scathing about the Phillips Inquiry calling it '£27 million misspent'. 'But you can't blame Phillips', argues Grant 'because he is not a scientist. But I look back on that public inquiry with dismay. If very distinguished people with all that mass of information can get it so wrong…'

Grant believes that Phillips is wrong when he says that scrapie does not affect humans. 'What we know is that we do not know whether scrapie affects humans,' she says. 'Look at it another way round – where in the wide world did that infectious spongiform encephalopathy, known as CJD, where did it come from before BSE? Well there is only one reservoir, scrapie has been with us since 1732. It's all over the world. I have always maintained that those people who are genetically susceptible and who eat the spinal cord fragment out of a lamb chop twice a week for twenty years are very likely to develop classical CJD. Prove I'm wrong. You can't. Mr Phillips is wrong.'

'If people had listened to what I was saying, and checked up on the references that I had provided, for example the work done on transmitting scrapie so easily to a long list of mammals, including apes, which are primates, by feeding it as an ordinary part of their diet. If they had paid attention, any intelligent person would have realised what it meant. They didn't listen. There would have been fewer cases of vCJD, because until November 1989 and even for some time afterwards because it wasn't properly infected cattle brains went into our meat products. That is because there was no way of knowing when the animals were going through our abattoirs, which of them had got infection.'

'In so much as the inspection of our food factories is accurate, it is safe now' argues Grant, 'but I do not trust them. The safety of our meat products depends entirely on the implementation of the Specified Bovine Offal ban. My experience, talking to people and abattoir people is that there is a shortage of the people who do the inspecting. If there is a shortage, the likelihood is they are not covering it all.'

Professor Richard Lacey

Another person who had been alarmed by Holt's BMJ article was Professor Richard Lacey, who had already made the headlines over listeria, chill-cook foods and microwaves. A Professor of Clinical Microbiology at Leeds University, Lacey became a key government critic of the BSE crisis. 'I was in a unique position, because I didn't need money from industry. In particular, I didn't need it from the food industry. Therefore I was able to say what I wanted. It hadn't occurred to me that my job would be at risk,' he recalls sitting in his dining room, surrounded by jigsaws of his garden, which he makes from his own photographs.

Lacey was the government insider who became the outsider. He had been a consultant to WHO since 1984 and had sat on the influential MAFF Veterinary Products Committee, but had become so disillusioned with the committee, he dismissed it as nothing more than a vehicle to 'rubber stamp government action.'[29] On the committee he maintains that they were fed 'selective information and summaries' and argues that the committee were 'very emphatically manipulated'. He confirms that on at least one occasion, minutes of a meeting were written before the meeting even took place.

By speaking out on BSE, he 'hammered the final nail in his professional coffin'. It took a personal toll as well, as Lacey and his family received death threats and his children suffered at school because of Lacey's outspokenness.[30] 'Mind you', he says with a wry smile, 'if I had my time again, I would have done exactly the same thing'.

Lacey was labelled 'hysterical and alarmist' by a government that set out to 'shoot the messenger'. He was ridiculed by politicians who used parliamentary privilege to 'call on the Department of Health to investigate' his mental state. Nine members of the Department of Animal Physiology and Nutrition at Leeds, who received funding from MAFF and food companies, argued that Lacey was effectively 'putting "putative" risks to the public before commercial and political concerns'.[31]

MAFF contemplated launching 'an all out attack' on Lacey 'at every opportunity particularly with respect to his published musings'. The officials even discussed monitoring Lacey's activities because 'knowing the enemy is the first step in developing a sound defence'.[32] It lead to Lacey being forced to 'give up my tenure at Leeds University'. He was told that his job was no longer needed as his department was merging with the Public Health Laboratory Service, forcing him to resign. Days later the decision was reversed, but after Lacey had retired. Lacey maintains that the whole campaign to attack him cost the NHS £500,000 in legal costs and pension rights.[33]

Lacey became subject to the 'Star Chamber' of the House of Commons Select Committee on Agriculture in 1990.[34] Lacey and Stephen Dealler produced a detailed paper for the Select Committee, called 'The Risk to Man', but the committee attacked Lacey's evidence as 'speculation, supposition, and conjecture'.[35] Lacey said that the politicians' attitude was to 'prove that something is dangerous, rather than prove that it is safe'.

When the Select Committee issued its report saying that UK beef was safe, Lacey was dismissed as being 'sensationalist' and as having 'lost touch with the real world'. His evidence was discarded as a mixture of 'science and science fiction'. The committee allowed the then Agriculture Minister, John Gummer, to tell the House of Commons that 'British beef can continue to be eaten safely by anyone, adults and children'.[36] The select committee was used to try and destroy a key government critic on food safety, whilst the government reassured people that there was nothing to worry about. The same tactic was again used years later against critics of GM.

But attempts to stifle Lacey were torpedoed by the Dorrell announcement on 20 March 1996 linking the new form of CJD to BSE. Even since then, the authorities have not taken BSE seriously, Lacey believes, and he cites the lack of comprehensive testing for BSE as a fundamental example of this. He also argues that nothing has changed. 'We have had a new government – a new approach to the whole thing, without any serious attempt to get rid of it, thinking that they will get away with it until the next election, and they probably will.'

'For many years now there is a simple test that can be done after slaughter on the brains of cattle to identify which cows are infected and which aren't,' he continues. 'If you have an infected herd, it should be destroyed and this happens in other countries. They should have done this.'

The professor was pleased with Labour's decision to hold a public inquiry into BSE. After giving his evidence he recalls how he nearly danced out of the hall, thinking that Phillips was 'a man who seemed anxious to collect all the known facts without political fear or favour'.[37] His optimism was short lived. 'I think the Phillips Inquiry was a total disaster', he says now. 'They took notice of people with zany theories and ignored the fundamental point which I tried to make. That is that sheep scrapie has been endemic for at least two centuries before they were exposed to feed. There was no rendering at this time. There is all the experimental evidence that it can be transferred vertically. If you have a disease that persists for centuries it has to be passed from animal to animal. Vertical and horizontal. Otherwise it would go away. I kept making this point and they kept wanting to believe the lies from the ministry. It is still going on. And SEAC seem to believe this too.'

Lacey maintains that the British government has hard lessons to learn. 'Still today they have not taken any significant action at herd level. Things are just as bad. BSE can be sorted out,' he says, 'by regular testing and progressively over decades eliminating the affected herds and restocking on new areas. If you restock on the same ground it could be reinfected.'

Even though Lacey was scathing about Phillips, the Inquiry called Lacey and Dealler serious scientists 'motivated by very real concerns about the hazards posed by BSE. Some of the conclusions they reached were speculative, some were extreme and some have been proved wrong. Many have, however, been vindicated as knowledge about BSE has increased. The public concerns raised by media coverage of views expressed by Professor Lacey were unwelcome to MAFF. With hindsight it seems to us that they were beneficial.'[38]

Professor Lacey believes it is still far too early to predict how many people will die of vCJD. 'Let's say the incubation period in cattle is five years, in mice eighteen months and sheep two and a half years', he says. 'In almost any human disease we are talking about a third to half the lifespan. You would expect the incubation period to be about thirty years, therefore if the exposure was greatest in 1988, 1989, which I doubt, you are talking about 2020' argues Lacey. But he adds 'I don't think you can predict how many cases there will be, the information is not available on which to make an assessment. We don't know what the infected dose is, we don't know how it is acquired, we don't know when it was acquired, we don't know if it is vertical or not. None of these are reasonable to experiment in people'.

Dr Stephen Dealler

Working with Professor Lacey at Leeds was Dr Stephen Dealler, an experienced microbiologist, who has worked in 35 hospitals in nine countries. He too became involved in the BSE crisis, after reading the 'exceedingly misleading' failures of the Southwood Report.[39]

Dealler had worked with Professor Lacey on issues such as listeria, microwaves, food irradiation and salmonella and had already come across the tendency of MAFF to 'deny everything'. Dealler argues that the raison d'être for MAFF goes back to its inception. 'It was all due to feeding the population after the war. I realised that the way MAFF worked was that they had to make sure that British agriculture made money,' says Dealler. 'Their job was not to create safe food at all. It was an economic one.' This meant that MAFF were 'telling a complete load of lies' and 'crossing their fingers and hoping it would just go away'.

Researching a scientific paper on BSE with Lacey, Dealler had to visit the leading TSE researchers around the world. Like Dr Holt he visited Hugh Fraser at the NPU, who explained 'how inadequate research was being carried out under directives from CVL. He knew what research was needed and was not being permitted to do it'. 'They weren't allowed to carry out research into BSE,' says Dealler, 'MAFF were sending them Directives telling them what to do. It shocked them'. The scientists also told Dealler they had stopped feeding beef to their families.

Dealler then flew to meet Richard Marsh, the American mink expert. 'He explained why one did not give up on BSE as being just scrapie, why the risk to humans certainly could not be pushed to one side and why one should assume BSE to be infective to humans until there was evidence that it was not.'[40]

Dealler did not believe the government reassurances that the risk of BSE passing to humans was low. His own estimates in the early 1990s showed that 'approximately seven cattle were eaten with BSE infection for every one that died with symptoms'. Learning more about the disease, he established that 'in other species muscle and other meats that we were continuing to eat had been shown to be infective'.

Dealler also found out that MAFF was stopping researchers at home and abroad from carrying out research into BSE. Even worse 'MAFF had not been carrying out the research, but they were telling people they had. A lot of the quite obvious and major research had not been done'. Dealler argues that the most obvious research was to 'inject a whole lot of bovine tissue into mouse brains', to see if it was infectious. Indeed this is what the scientists at the NPU in Edinburgh, wanted to do. They had started working on experiments but 'they were told to stop'.

Lacey and Dealler also found, to their cost, that MAFF 'owned all the cattle that might have BSE'. This meant that no research teams could get BSE infected tissue. 'They did not want us to have any of this material.' When Lacey and Dealler managed to get hold of a cow born after the cattle feed ban, the duo had it slaughtered and analysed the brain. 'We showed it had BSE. But MAFF was denying that any cows born after that date would have BSE. We published it saying "yes they did have BSE and here is one"'. Dealler was phoned the next day by MAFF and told to give the samples back. 'That cow was a major risk to MAFF because I could have given that tissue to someone to inoculate mice brains with.'

Then in May 1990, public concern exploded when a cat – dubbed Mad Max – was diagnosed as having BSE. The disease had jumped another species. But still MAFF denied any risk to humans. 'Cats do not get infected with scrapie,' says Dealler. 'But they were saying that if a cat had caught BSE it did not matter. But they had also been saying that humans wouldn't go down with BSE because humans did not go down with scrapie. What MAFF was saying was obviously dribble – it obviously wasn't the same as scrapie – it was different.'

Dealler and Lacey's paper entitled 'Threat of BSE to Man', was published in *Food Microbiology* in 1990. It concluded 'the high prevalence of infected animals must represent a phenomenal danger to man', but was dismissed as 'speculation and scaremongering' by MAFF. Their arguments had to be 'knocked down', argued MAFF's chief veterinary officer. The BSE Inquiry found that this 'careful' article did not deserve the 'opprobrium' it received from MAFF.[41]

Professor Lacey and Dr Dealler met SEAC for a private meeting in June 1993. Dealler had calculated that everyone eating beef at that time has eaten about 50 infected meals. 'Lacey and I were being treated as people who simply did not understand the subject and we had to be taught the truth'. More worrying for Dealler was that he felt that SEAC – the government advisors – were 'accepting MAFF's position on BSE as fact.'[42]

Meanwhile MAFF had gone on the offensive. 'They had rubbished Lacey to a point where the media would not talk to him. Helen Grant had been out the field for quite some time. They had very much rubbished Harash Narang. They were out to get him. There was only me left. I was the only one,' says Dealler.

Dealler was attacked too. 'I was really, really dumbed down by the medical profession' he says, recounting a 'disgusting' episode that happened. 'By 1993, I had been applying for consultant posts'. Over-qualified for many of the positions he was applying for, he thought he would get a job. 'I applied for 20 posts. To start with I got interviewed. Then I didn't even get an interview. Eventually someone said to me "Look Steve they are doing you down. You are not going to get a job". "I said to myself, who is doing it?"'

Dealler applied for a job at Burnley Hospital, where he was already working. 'They realised that I was perfectly good and even over qualified for the hospital. But when I applied for this one someone from my department came up and said "Steve – what's all this – we have received a letter about you. Apparently you are grossly incompetent and should not have been given a consultant post". He showed me the letter.' It was from someone from the PHLS. 'The only trouble was – they [Burnley] knew I wasn't.'

This person from the PHLS was on the selection committee for the Burnley position. 'After the interview he said "it's quite clear we won't be considering Dr Dealler, will we?" but it turned out I was way ahead of all the other applicants.' Dealler came to the conclusion that it was a 'central thing' that the letter had been sent. 'I presume he was told to say this. The problem is that they could find no way to stop me.' Even years later, Dealler comes across effects of the smear.

By 1994, Dealler published his work showing that people were eating large numbers of infected animals. It was denied by MAFF. The same year when he sought a meeting with the new Minister of Agriculture, Mr William Waldegrave – it was turned down.[43] Dealler was convinced by now that SEAC was acting 'unethically'.[44]

But says Dealler, 'I was the only person in the UK looking at methods of treatment. I went to the Public Health people and told them that we must do this and they said that "No we have been told not to – all work on BSE had to be done by MAFF – no work can be permitted to be done by Public Health". That had come as a directive from the Department of Health.'

The scientific journals, too, were afraid to publish 'anything that was potentially risking them, they would not dare'. Dealler argues that this is still happening and he cannot get a paper published talking about the risks of contracting vCJD from dentistry. The paper argues that the risk is low, but can be altered by positive measures. He believes that dentistry is one of the main risks of continuing exposure to BSE. 'The risk of eating beef has reduced dramatically', he says. 'Anyone over the age of five in the UK has been exposed to BSE to such a degree that the added exposure from now on is minor. Whereas with dentistry this may not be true'.

But it is not all bad news. In 2001, Dealler noted how the 'good news' is starting to come through. 'There are now over 40 drugs that are active against prion infection and we have some that are of adequately low toxicity to permit testing on patients' he wrote. Dealler said that some drugs might also help Alzheimer's.[45]

Dealler continues to look for solutions. 'I have been working on treatment and diagnosis of vCJD and the diagnosis system will come through and the treatment will come through' he believes, but says he is still hampered in his work. 'What we are trying to do is produce a test and treatment for the person who has no symptoms. So that we can then decide who gives blood. We are going to have to test everybody. You have to treat them before you have symptoms. It would be unethical not to test.'

Like Lacey and others he believes we still do not know the full extent of the vCJD epidemic. 'We are going to see the incubation period between 15 and 40 years, with a peak about 25 years. If you try to work out when you would expect to get the peak it would be between 2010 and 2020.' Dealler believes that if the incubation period is until 2010, there will only be 1000 cases of vCJD, but 'if it's 2020, it's in millions; we don't know'.

With unknown number of people unwittingly carrying an incurable disease, it seems logical to start testing blood donors. But when a test does come onto the market, it is estimated that about half of the UK's donors might stop giving blood rather than be told they have an incurable disease.

Dr Harash Narang

Whilst Stephen Dealler's team carries on working towards a live test for vCJD/BSE, there is one scientist who believes that he has one that works and that it could have altered the whole BSE epidemic. But like all the other scientists, Dr Harash Narang was marginalized by MAFF. He was also sacked from his government job. 'His life has been ruined by the Ministry of Agriculture', says Dr Helen Grant. Stephen Dealler adds: 'They were determined to nail Harash Narang into the ground because he said we should not assume that it isn't a risk to humans and in the UK they were determined to believe that it was not a risk to humans.'

Dr Narang, who is a microbiologist, has over 30 years experience working with TSEs, examining his first human victim of CJD as far back as 1970. Most of this work has been into slow viruses in the UK and USA where he has worked with Carleton Gajdusek, the Nobel Prize winner. Gajdusek has called Narang's work brilliant and important.[46] 'I started work in 1970 on this disease. Now in 1988 I pointed out that there was a problem and you name me one thing where I have been wrong', he says.

Dr Narang's controversial contribution to the BSE debate stems from his development of diagnostic tests for BSE and CJD. 'The last thing in the world they wanted was some clever chap with a test to reveal that the vast majority of animals going through the abattoirs, although they looked lovely from the outside, were actually infective', argues Helen Grant. 'Everything MAFF could think of to throw at Narang, including cloak and dagger exercises, he had it thrown at him.'

Narang was sacked from his job, and sidelined by the medical establishment. His flat has been broken into, where people were only 'interested in documents' and 'removed files' and he says his car has been tampered with. This was done to frighten him, Narang believes.[47]

By the late 1980s Narang says he saw a 'change in the pattern' of the CJD brains he was analysing. Both the number of cases increased as did the distribution and 'clinical presentation' of lesions in the brains.[48] He started to inject animals with the brain of the CJD victims, the only scientist to do such work. 'By 1990,' maintains Narang, 'these experiments were well advanced and were giving clear preliminary indications that they were CJD cases infected with the BSE strain'.

Dr Narang maintains that the PHLS had the animals destroyed before the results could be worked out. He argues that 'had these experiments been completed in their entirety and had the preliminary indications been confirmed we would have been in no doubt about the link between BSE and CJD and many lives could have been saved'.[49] Narang's version of events is challenged by the PHLS, who deny that the animal house was closed down and question Narang's tests.[50]

In March 1990 Narang and a colleague, Dr Robert Perry, a consultant neuropathologist at the Institute of Pathology from Newcastle General hospital, submitted a paper to *The Lancet* saying that they had recorded increased numbers of CJD cases in the Northern Health region.[51] Dr Robert Perry, too, was subsequently victimized and branded by MAFF as a 'loose cannon who could cause mayhem'.[52] The experiments were finally started six years later by MAFF. 'The government has not just misled the public on BSE, they have deliberately sabotaged science while at the same time pretending to be objective and scientific,' argues Narang.[53]

Narang was also working on diagnostic tests for both BSE and CJD. The first test – known as a touch test – could only be used on dead animals, the second – the western blot – could be used on both live and dead and the third – a urine test – was exclusively to test live animals. Dr Narang had developed the touch test by the late 1980s. It was a same-day test, as opposed to the slow laborious one used by MAFF. 'The touch test was at the time that animals were being killed in abattoirs and you wanted to know that they were positive or negative, and know that quite quickly. It is not that it is impractical to test every animal,' says Narang.

Narang believed he could test 20–30 animals a day at the abattoirs. But the government did not want to know the answers, says Narang. 'Harash, you know these animals would be positive' he was told by a senior MAFF official. 'Mr Gummer knows they would be positive. He doesn't want you to rubber stamp it.'[54]

Indeed, internal MAFF correspondence shows that there were 'political sensitivities in producing a diagnostic test for BSE and the pressures this may bring eg for major screening programmes.'[55] In January 1990, MAFF turned down Narang's funding application, after government scientists claimed that his test was 'extremely insensitive … for the diagnosis of clinical BSE'.[56]

At the time, the principal opposition spokesperson on farming was David Clark, who believes that MAFF 'systematically set out to scupper' Narang's work, because, if Narang were 'successful in his diagnostic tests the whole of the British beef industry would be put at threat'.[57] So that was the bottom line – put the UK herd at risk by testing, or put the UK public at risk by not.

By March 1991, Narang's career started to nose-dive, not an unknown occurrence if you questioned the government's line on BSE. His experimental work was stopped 'for reasons of safety' and he was charged with working with unlawful chemicals by the Health and Safety Executive (HSE). The following year he was suspended from the PHLS for 'professional misconduct', having supposedly breached medical confidentiality with patients' records. His life became a 'nightmare'.[58]

By 1993, Narang had developed a 'live' urine test to test patients with CJD, based on the western blot techniques. The equipment cost is about £20,000 and Narang says that in the size of a small room, you could test half a million samples a day. 'It is quick and speedy,' he says but 'all they keep on saying is that we will give you a chance' but in reality 'they just want to wear you down'.

In November 1994 Narang was finally made redundant by the PHLS who say they have subsequently been vindicated as Narang lost an industrial tribunal he brought against them. They also say that other researchers did not validate his tests.[59] In contrast Harash says that the then Minister for Health, Stephen Dorrell, told the 'PHLS not to be involved with BSE/CJD. The PHLS had orders from Stephen Dorrell to make me redundant, and not to be involved or seen to be involved'.

Relatives of vCJD victims argue that Narang's test works. As their son lay dying, Narang was asked by the parents of Peter Hall to undertake a urine test on their son, after being repeatedly told that Peter was too young to be suffering from CJD. Peter's mother, Frances, believes that 'both the urine test and subsequent post-mortem not only confirmed that Peter had, in fact, died from vCJD but left the government with no alternative but to admit that a new strain of CJD had appeared.'[60]

Narang believes things could have been so different. 'Had any of these various tests been funded or supported in any way by government bodies', argues Narang, 'it would have been possible to set up a slaughter house test for diagnosing subclinical BSE in cattle that had not yet shown the symptoms of the disease, but which were entering the national diet'. If this had happened, Narang maintains 'the disease could have been eradicated' and the 'fundamentally flawed "culling policy" could have been avoided'.[61]

Today Narang is still looking for buyers for his diagnostic test. He also believes that history may be repeating itself, this time with the risk of CJD from blood transfusions. Narang maintains that although the CJD caused by BSE is clinically similar to the CJD caused by 'human to human transmission,' he can tell the difference between the two by looking at brain damage. At the end of 2001 he published the results in the *Journal of the Society for Experimental Biology and Medicine*, arguing that the distribution of 'PrP plaques in the brain of the patient strongly suggests that this case was infected through blood transfusion'.[62]

But the government only acknowledges a 'theoretical' risk through blood transfusion. Government scientists argue that 'there is no epidemiological evidence to indicate that iatrogenic CJD has ever occurred via blood or blood products', which angers Narang. 'They say there is no case of CJD by blood transfusion, but there is a case' he retorts. 'I have identified a case of CJD who had died after having a blood transfusion and who had the hallmark of being infected by the blood transfusion.'

However, the authorities are worried. In August 2002, the Department of Health announced that it would be sourcing blood for transfusions for young people and babies from the USA. But it still argued that the risk remained theoretical.[63] Once again the authorities were denying the evidence other scientists said existed. Narang subsequently warned of the risk of bringing in 'untold other diseases' from the USA, where the screening practices are different. He also believes that 'these are the same old statements' that we saw on BSE and how the government will go from saying 'absolutely no risk', to there is 'no risk' to, 'oh, there is a risk'.

You could argue that the government is not keen to introduce a screening programme in that would reveal the scale of those incubating vCJD. In 1998, Dr Narang told Phillips: 'I would like to utilise this inquiry for the purpose of a plea that all future blood donors be tested utilising my urine test to prevent further transmission from blood products'.[64] This has still not been done.

Mark Purdey

Another critic of the government's handling of the BSE crisis is not a scientist but an organic farmer, Mark Purdey, who has also suffered vilification and harassment. His views remain highly controversial, as he believes that the BSE epidemic is not due to the recycling of MBM, but is due to organophosphates (OPs).

'It all started in 1984 when a government officer asked us to use this warble fly treatment', says Purdey. 'He said that we were in a compulsory zone and had to comply. We refused. My grounds were that OPs are derived from military nerve gases.' Purdey believed that it would cause an epidemic in neurological disease in cattle to pour Phosmet along the spine just millimetres from the central nerves. 'I said it would also damage the farmer. We prevented MAFF from forcibly treating our cows in the high court through a judicial review, so we won the day.'

'When BSE arrived – I thought – here we go – here is the boomerang effect of this chemical treatment. It was known that TSEs were caused by a malformed prion protein. So I argued right from the beginning that OPs could be causing BSE by deforming this new protein by deforming its molecular shape, just as it is known to deform other proteins in the nervous system.'

In 1988, Purdey wrote to *Farmers Weekly* saying that BSE could be caused by the direct application of systemic Phosmet, the trade name for the OPs, which produced a delayed neurotoxic effect and had caused BSE. 'Even though my theory has evolved today to bring in other factors, that cornerstone of my theory I still hold today', says Purdey.

Phosmet had been used in the UK since the 1970s for control of the warble fly. Britain, Purdey says, was unique in allowing a particularly high dose – 20 per cent – of Phosmet concentration, which was granted in November 1978. 'This is what everyone was using. By the 1980s there were other types of OPs which had been phased out. It was only Phosmet that you could get that locked up the copper'. This brings in the rest of Purdey's theory: 'My own funded research has found new factors. The actual mechanism of the OPs did not become apparent until I started doing this global tour'. Purdey travelled to places such as Colorado, Iceland, Slovakia and Japan – areas where scrapie or CJD was prevalent. What he found was that the Phosmet type of OP actually captured copper in the nervous system. 'I visited the clusters of the traditional disease. In all of these clusters I found a common abnormal mineral template, which was low copper, and high manganese or silver in the environment. ... When there is a lack of copper in the environment, then you get susceptibility to disease, and if you get a high loading of a metal that will substitute on that site on that prion protein such as silver or manganese. In Iceland I found that in valleys where scrapie occurred had high manganese and very low copper. Where it didn't occur the levels were normal. It was like black and white.'

Whilst Purdey believes BSE has declined due to the reduced use of Phosmet, he also feels that the recycling of the infectious agent through MBM did not 'cause the disease. I always argue that what is most bizarre is that the MBM which is blamed for BSE was sold all over the world. To me that is the most major flaw in the theory', he says. Purdey asks why, if micro-doses caused 40,000 cases of BSE to be caused after the ruminant feed ban, why didn't the 'mega-doses' that went to the Middle East, Asia, South America and South Africa, cause epidemics there. 'You can hide a few cases, but they should have had an eruption. It should be world-wide.'

Many independent and government scientists, officials and politicians dismiss Purdey's work, saying that BSE is a transmissible infection, whereas OPs are neurotoxins. Dr Narang believes that there 'is no reliable link between the number of BSE cases and the use of OP substances', but concedes 'it is possible that neurotoxins do play an indirect role in reducing the incubation period, since some of these substances damage the nerve cells'.[65] The Phillips Inquiry argued that Purdey's OP theory was 'not viable, although there is a possibility that these can increase the susceptibility of cattle to BSE.'[66]

Professor Smith from SEAC believes Purdey's theory 'is not a consistent explanation for BSE and CJD, which hangs together less well than the MBM contamination theory.'[67] Dick Sibley, the President of the British Cattle Veterinary Association, says Purdey's theories are of 'academic interest'. But as we still have 21 cases of BSE that 'defy explanation', Sibley suggests it might be time to look at 'earlier theories'.[68]

Purdey does have supporters. Environmental columnist George Monbiot wrote that Purdey 'could be about to overturn the entire body of scientific research on the biggest public health scandal of modern times' and if he did so 'he deserves a Nobel Prize for medicine'.[69] Dr Steve Whatley, from the Department of Neuroscience, at the Institute of Psychiatry, London, examined Purdey's theory experimentally. He concluded that: 'whilst our results therefore provide no evidence for the hypothesis that Phosmet may directly cause emergence of BSE ... it represents the first experimental evidence that Phosmet may be capable of modifying risk of transmission'.[70]

Dr Joseph Forde Gracey, a former Home office veterinary adviser, says 'we cannot afford to ignore' Purdey's work 'because we do not know what the answer is'.[71] The government looked for answers when it asked Sir Gabriel Horn, Emeritus Professor of Zoology at the University of Cambridge to examine the causes of BSE. Although the committee was said to be 'independent', Horn is part of the scientific establishment and had been involved in the BSE saga before, chairing a working group in the early 1990s that apportioned BSE research money.[72]

Horn dismissed Purdey's theories, arguing that it cannot be 'reconciled with the epidemiology and is not supported by the research'.[73] The report enrages Purdey. Horn argues that 'nearly all cows which developed BSE were born after 1982 so were never treated with Phosmet to eradicate warble fly.'[74] Purdey calls this 'blatantly bogus' as warble eradication first became compulsory in 1982 and the government themselves designed the warble-fly eradication campaign. 'They of all people know it became compulsory to use the chemical twice a year in the autumn and in the spring – why did I have to have a high court case in 1985, to prevent their use, when it was not even used – it is nonsense. In 1995, we were being mailed by the government with leaflets saying please warble fly your cattle'.

Purdey also says that Professor Horn's committee noted that there was more BSE on Guernsey than Jersey, when Jersey was a compulsory warble fly zone and Guernsey was not.[75] 'Well Jersey was not a compulsory warble fly zone' retorts Purdey in anger. 'The only people who have ever said that are the chemical industry.' Purdey wanted to sue Horn for defamation. 'But they said it was qualified privilege.' However, Purdey's legal letters may have had an effect – the report has been dropped from the Department for Environment, Food and Rural Affairs (Defra) website.

Purdey cites the Horn report as another case where he believes the political and scientific establishment has been out to discredit and harass him. He also claims that his emails have been tampered with. 'The real weird things went on in the eighties though,' he says. 'In the early days it was actual physical intimidation which was a lot harder to handle.'

Purdey says he was harassed by a neighbour who suddenly started to victimize him after he took MAFF to court over OPs. This person shot at Purdey and terrorized the family, making Purdey's life 'hell'. Purdey believes his phone was tapped, and his rubbish searched. When Purdey went to the police they told him 'You realise that some people are employed to behave in this way? You were the one that took the Ministry to court over warble fly'.

The same week that Purdey sold his farm to move to Wales, the nuisance neighbour suddenly put his 'dream retirement home' on the market. Purdey's new Welsh home was burnt down before he moved in. The police said it was an electrical fault. Purdey believed it was arson. 'I suffered the price personally. It has put loads of pressure on my relationship. If you start talking about people firing guns at you people think you are paranoid. However much people say "oh that crackpot farmer Purdey", it doesn't worry me. The truth is the science and you can't run away from that.'[76]

'We haven't learnt from our mistakes', says Purdey, 'by suppressing the evidence on the original causes of BSE they are evading the issues again. Just like they are doing with GM foods. We haven't learnt a thing. All the government bodies are just looking after the commercial interests of the companies rather than the long-term health of the people'.

CHAPTER 4

Silent Spread

'Nearly all of the damaging effects to the people, environment and economy were caused not by the disease, but by the government's chosen strategy to control it'

Alan Beat, Devon small-holder[1]

'My advice wasn't rejected – it was never sought. I had two phone calls in the whole outbreak from the Government's Chief Scientist, David King'

Professor Fred Brown

If anyone thought that the government had learnt lessons from BSE they were wrong. When foot and mouth disease struck UK agriculture in 2001, the feeling of déjà vu became all too apparent. Dissenting scientists, who were world experts in their field, were once again marginalized, as a small group of elite scientists formulated policies that had a devastating impact. Macro-economic policies and trade concerns overruled those of common sense, precaution, animal welfare and health protection.

Professor Fred Brown

In the mid-1950s a young scientist called Fred Brown began working on foot and mouth disease at the Government's Pirbright Laboratory. Brown became an expert in infectious diseases, being asked to join Southwood's BSE committee in the late 1980s, which he declined as he felt he 'did not know enough about it'. There were people like Alan Dickinson and Hugh Fraser, from the NPU, who 'would be much better people to be on that committee'.[2]

Professor Brown sat on SEAC from 1990 until 1998, when he was 'rotated' off. A colleague on SEAC later recalled that 'there was a danger of us becoming simply a talking shop and an adjunct to MAFF upon whom we were essentially dependent. Against this, Fred Brown with his sharp mind, massive research and research management experience, independent spirit, and plain speech was a bulwark'.[3]

In the early 1990s, Brown had argued that the BSE problem was too great a job for a part-time committee, arguing there should be a full-time government advisor, but his request was turned down, and he was told by the head of SEAC that 'we couldn't think of anyone suitable'. Brown replied that it was rubbish. 'There are plenty of people who knew about these diseases, Alan Dickinson knew more than anybody.'

But foot and mouth disease would be Brown's forté. This is a highly infectious virus that affects cloven-hoofed animals including cattle, pigs, sheep and goats. Clinical signs include blisters on the mouth and feet, reduced milk yield, high temperature and lameness. The symptoms are easily recognized in cattle and pigs, but in sheep the virus can go undetected. The young and old are particularly vulnerable, but foot and mouth disease does not affect humans.[4]

Brown stayed at Pirbright for 27 years working on foot and mouth disease. Although the last serious outbreak in the UK was in 1967, there was a small outbreak on the Isle of Wight in 1981. The disease is endemic throughout two-thirds of the world. Between 1999 and 2001, there were some 14,898 cases of foot and mouth disease in 16 different countries.[5]

Brown was involved in the 1967 outbreak. 'The lesson that is learnt with foot and mouth disease' Brown says, 'is that it is so highly contagious, that you must stop movement'. Once that is in place, 'you kill infective animals and you also kill all the susceptible animals on that farm, because it could spread to them'.

In the late 1980s, Brown was invited by Roger Breeze, at the time Director of Plum Island Animal Disease Center, the 'Alcatraz for Animal Disease', in the USA to work with him on foot and mouth disease. He has been there ever since. With 47 years experience working on the disease, Brown is one of the world's leading authorities on the subject. He has published nearly 400 scientific papers, the majority on foot and mouth disease. 'My contributions will be trying to understand the immunology of it, what protects when you vaccinate,' says Brown now. 'The molecular issues that I have been involved in have led to a way to diagnose the disease, and alternative methods of diagnosis. This means that you could tell the difference between vaccinated and infected animals and how you could identify an infected animal by a laboratory test. You could then extrapolate that to the farm gate.'

In January 2001, collaboration started between Plum Island and the American company Tetracore, a biotechnology research and development laboratory that develops diagnostic reagents for infectious diseases and biological weapons.[6] 'Tetracore wanted, ahead of any foot and mouth disease outbreak to look at all the infectious agents on the famous OIE List A', says Brown. The OIE (Office International des Epizooties) was set up in 1924 to 'to guarantee the transparency of animal disease status world-wide'. Its List A includes diseases such as foot and mouth disease, classical swine fever as well as animal fevers, influenzas and poxes.

The whole point of the test was to get rapid results 'very, very quickly', said Brown. Speed is the essence in dealing with the disease. The faster you stop movement, the faster you diagnose, the less you have to kill. A fast and portable diagnostic test would be a real breakthrough in the way that authorities treated the disease. 'They had developed all these tests using the nucleic acid of the agent', Brown recalls. 'I was brought into this because of my foot and mouth knowledge and the reagents I had. We did it and very rapidly we said that it was a good test.'

Work was continuing on the test when on 19 February 2001, foot and mouth disease was suspected at an abattoir in Essex and confirmed the following day. 'When the outbreak started we concentrated the effort a bit and we said it works for all the seven sera types of the virus,' says Brown. Of the seven strains, the UK epidemic was caused by the PanAsia strain of foot and mouth disease type O virus. At the time of the outbreak there was wide speculation on how the disease entered the UK, either through imported infected illegal meat or another route. The outbreak at the abattoir in Essex was officially traced back to Burnside Farm, Heddon-on-the-Wall, Northumberland, where it is now thought that the pigs had been infected since 12 February and that it 'was probably present at the beginning of February/late January'.[7]

Rumours continue that this was neither the date, nor the location for the start of the outbreak, and that the farmer concerned, Bobby Waugh, was made the public scapegoat.[8]

The official line, that the pigs ate contaminated pigswill, became embarrassing for Ministers because of revelations that SEAC had argued for the banning of pigswill from catering waste in 1998 to prevent disease transmission. The recommendation was rejected at the time on grounds of the economic consequences.[9]

While Brown and his colleagues worked on their test, the UK response was slow and unprepared, despite warnings that a foot and mouth disease epidemic was imminent.[10] Six months earlier, an Irish vet had told a scientific conference how Europe was certain to be hit by foot and mouth disease. Millions of animals were likely to die, due to a combination of intensified farming, increased animal movements and a virulent new strain of the disease called PanAsia type O strain. Another scientist, a researcher from Pirbright, called the disease a 'major threat'.[11]

Indeed the scientists at Pirbright, designated the World Reference Laboratory for foot and mouth disease, had been watching the spread of PanAsia with increasing alarm. The head of Pirbright, Dr Alex Donaldson, says: 'When you have countries such as Japan, which has been free since 1908, Korea free of foot and mouth disease since 1934, South Africa never having had type 'O' up to the year 2000, all suddenly becoming infected with foot and mouth disease, it's an indication that there are different trends. Now that could be due to changes in legal trade or more illegal trade, but I think those episodes were indicators of increased risk'.[12]

But the UK was hopelessly unprepared, and one of the main tools to control the disease around the world – vaccination – was ruled out for political and trade reasons. Controversial tactics never employed before, such as a 3km contiguous cull, were used. The individuals in charge of the government's response were guided by flawed computer modelling that had never been subject to peer review. This means that the number of animals killed was on a scale previously unseen. The countryside was effectively closed, whilst rural businesses went to the wall. Farmers watched as riot police helped slaughter teams kill their animals.

The epidemic lasted for 32 long, painful weeks, the last case being confirmed on 30 September 2001 on a farm in Cumbria. But it was not until January 2002 that the UK was re-instated as 'free of foot and mouth', and it was only in September 2002 that the first UK beef was exported to Europe again.[13] This was a staggering 19 months after the beginning of the epidemic.

By the time the disease had been eradicated, 'more than six million animals had been slaughtered: over four million for disease control purposes and over two million for welfare reasons' concluded the National Audit Office. 'The direct cost to the public sector is estimated at over £3 billion and the cost to the private sector is estimated at over £5 billion.'[14]

But there are those who dispute these official figures, saying they were fudged. In official calculations, a breeding sow with six piglets would all be counted as 'one unit', even though seven were killed. The Meat and Livestock Commission estimate that, including all lambs, piglets and calves, some 11 million animals were killed. Their breakdown is approximately 9.5 million sheep, 860,000 cattle and 430,000 pigs.[15] After the epidemic was over, scientists came to the conclusion that cattle and larger farms were more likely to have transmitted the disease, rather than sheep and small farms. Yet many more sheep were slaughtered.[16]

A conservative figure is that the country lost £8 billion – other estimates say the final figure could be as high as £20 billion.[17] To protect what exactly? The export market affected by foot and mouth disease amounts to around £310 million. Even before the absurdity of the economics becomes apparent, these figures hide the continued misery that shattered lives across the UK as farmers watched prized livestock, some bred for generations, wiped out.

The images of burning carcasses would freeze deeply into the memories of millions. 'If you sat down and thought of the worst nightmare possible for farming this would be it', said the National Farmers' Union South West director Anthony Gibson. 'It is horrendous. This is just piling on the agony.'[18]

In contrast, when Fred Brown first heard about the 2001 UK outbreak, he told his colleagues that the UK authorities would soon have everything under control. They knew how to treat the disease, he thought. 'The great mistake that was made was the time they took to confirm it was foot and mouth. It took thirty hours before they said "yes this really is foot and mouth".'

It took three days for the government to ban livestock movements. In that time there was a massive movement of animals, resulting in the government losing control. Professor Mark Woolhouse, a government advisor on foot and mouth disease, believes that the single act of not stopping livestock immediately increased the size of the epidemic by up to 50 per cent. There were over a dozen key movements of infected animals that spread the disease from Cumbria to Anglesey, West Yorkshire and North Yorkshire and from Devon to Wiltshire and Northamptonshire.[19]

The government, and MAFF in particular, were also accused of failing 'to understand the way the modern livestock industry operates'.[20] MAFF did not even have an accurate list of farms or farmers or which of the 121,000 farms even held livestock. Its contingency plan for dealing with foot and mouth disease was hopelessly out of date.[21]

In the three weeks when it is suspected that the disease was undetected, MAFF thought that 2000 sheep movements had taken place, when the figure was 1.35 million. This movement of sheep was one of the main reasons for the epidemic. It appeared that MAFF did not realize that sheep, pigs and cattle were being moved up and down the country to be traded and slaughtered.[22]

The outbreak demonstrated to a bewildered public the long distance nature of industrial farming, an issue that animal welfare groups, such as Compassion in World Farming, had been campaigning on for years. In the years since the last outbreak, the number of markets had decreased from 380 to 180, but the number of abattoirs had gone from 2200 to

360.[23] There were 60 per cent more sheep in the UK in 2001 than in 1967 and many were destined for export. Another ironic factor was that the UK export markets, especially that for lamb, were seen to be buoyant at the time, profiting from the European BSE crisis – a crisis exported from the UK in the first place.[24]

'I know what I would do', Brown told his boss, 'I would put the army on every farm gate and stop movement until the farm was sure the animals were clean and could be moved. I would have then tested those animals rapidly. We had this test ahead of the outbreak. So the army on every farm gate, and I would have tested the animals, but I wouldn't have stopped them earning a living.' If the animals were free of the disease, says Brown, then they would have been moved. 'Instead they messed around. The Prime Minister brought in all these modellers to say what should be done. They modelled on the basis of what Donaldson had said in 1981, that the virus had moved 170 miles from France to the Isle of Wight. They based their models on the fact that this virus was airborne. In fact subsequent experiments at Pirbright by Donaldson showed it barely moved.'

Government advisors came up with the idea of a 'fire-break' in the form of a contiguous cull, described by the *Western Morning News* as 'crude, medieval and extremely brutal'.[25] 'That is what took the resources, because if you are killing infected animals that is one thing, if you are killing animals on an infected farm that is one thing, but killing animals within a circle of 3km, that becomes an enormous task. It had never been done before', says Brown.

On 9 March, just over two weeks after the 'official' start to the disease, Roger Breeze, now assistant to the head of the USDA in the USA sent an email to Alex Donaldson at Pirbright. He copied the email to Jim Scudamore, the Chief Veterinary Officer. Plum Island was offering the Pirbright scientists Brown's test, a 'real-time PCR' Smart-Cycler test that could identify which animals were infected with foot and mouth disease. Crucially, it was a mobile test. 'So we offered the test to Pirbright – they said they were overwhelmed. I thought the reaction was what it would be,' says Brown. 'I had warned Roger Breeze before he sent his email. I told him that there was no way you would get collaboration. But it's a pity. The nuts and bolts of how to do tests and how to arrange the samples was very easy. Any competent lab person could have done that.'

Brown is adamant the test worked, could have been put in a mobile van and would have told farmers within two hours whether their livestock was infected. 'We had done this experiment with lab specimens and we had shown that we could detect about 100 particles of the virus. A lesion or a blister will have about 10 billion particles. That was the whole thing. Speed is of the essence.' Instead of farm gate testing, the samples were being sent to Pirbright to be analysed, a process that took time, but Brown argues the two tests could have been run together and compared. 'That's how I would have done it, if I had been in charge.'

The real reason why the test was rejected, says Brown, was essentially 'political'. 'The thing the test would have done, particularly in the three-km killing, they call it cull, because it sounds nicer, it would have told people that they weren't infected. There are many, many stories of vets going onto farms, they would see a blister, and say foot and mouth and kill them, and when the sample went to Pirbright the test was negative. But by that time the animals were dead, and that's no good. I think it has angered a lot of people. There are many instances where the samples were negative and therefore those animals should not have been killed. They could have avoided the 3-km cull – that is for sure.'

In the middle of May, Brown and two experienced colleagues from Tetracore went with the head of the US Agricultural Research Service to see David King, the UK's Chief Government Scientist. 'I just described the results we had got. When the test had been turned down, we had done some experiments at Plum Island, where one of each of the three main species – cattle, pigs and sheep – one of each of those was infected. Then four of the same species were put into the same room. So it was infection by contact... We found we could detect the virus, sometimes a day or so before they showed any signs of the disease at all. That was good. Then came all the issues of false negatives and false positives. But that is exactly what we wanted it to do. It was an opportunity missed. It could have been arranged, I'm sure.'

But Brown and his colleagues knew that some sections of the scientific community would treat their test with scepticism until it was published in a peer-reviewed journal. So they submitted a paper on the PCR test for publication to the *Veterinary Record*, the weekly official journal of the British Veterinary Society. It was rejected, 'on no serious scientific grounds', says Brown, who also submitted two smaller items of correspondence on testing and foot and mouth disease, both of which were rejected. 'It was political,' he argues. Although the *Veterinary Record* refuses to disclose why the paper was rejected, the same paper was finally published, unchanged, in June 2002 in the *Journal of the American Veterinary Medical Association*.[26]

Whilst Brown's test was widely reported, what was not was the collaboration between Pirbright and the Dutch health care company, Intervet, a subsidiary of Akzo Nobel, the animal health care company. Intervet had been cooperating with Pirbright for a number of years. 'Intervet were collaborating with Pirbright on a rapid test using serology, not using PCR, nothing to do with nucleic acid. It is less sensitive, although valuable' says Brown. The test also takes longer, argues Brown, who says that another company – based in Kent – offered Pirbright a test, but this was also turned down due to the collaboration with Intervet.

Brown had already found out to his cost that the Pirbright–Intervet commercial relationship precluded Pirbright from cooperating. 'I think they were very busy', says Brown, but he believes other commercial and political factors played a part. Back in the late 1990s, whilst working on a serological test, Brown had asked the lab at Pirbright and other labs in Argentina and Holland whether he could use some of their serum that they had supplied in the past to the Plum Institute. 'Pirbright turned down my request'. Asked why, Brown responds that: 'They were collaborating with Intervet – they didn't say that though, but they had a collaboration. That was the answer I got two or three years before this current outbreak, and that is why I sensed it would happen again and it did.'

Just three weeks after the scientists at Plum had contacted Pirbright, Intervet put out a press release announcing it was working on a new serological test as a veterinary breakthrough. When completed it would make the 'identification of infected animals possible in a population where routine vaccination is used to prevent the clinical symptoms and spread of the disease'. The press release continued that 'in this way the test offers opportunities for an effective eradication policy with the help of vaccination'.[27]

In October, Intervet issued a further press release announcing the test was ready for use. 'The marker test, combined with modern foot and mouth vaccines, enables authorities to use vaccination in foot and mouth control strategies and avoid massive preventive culling of healthy animals.'[28]

The official version as to why the test was turned down is different. The Chief Scientist, David King, maintains that Brown's machine was 'untested' and there are 'very serious questions to be asked about the use of that machine in the field, in particular the problem of cross-contamination'.[29] Dr Alex Donaldson at Pirbright was undoubtedly overwhelmed, but adds that the normal scientific approach is for data from these tests to be made 'available through the scientific press' so that one 'can start looking at that data in relation to well-established gold standard methods'.[30]

So the diagnostic test that Fred Brown was advocating – although tried and tested to work in the lab – was untested in the field and non-peer-reviewed. This was the official reason it was turned down, although we now know that a commercial tie-up between Pirbright and Intervet prevented its use. Brown argues that if his test had been used, this would have prevented the contiguous cull – the slaughter of some 2 million animals.

Yet the government was just about to embark on the contiguous cull – based on a computer model that was also untested, non-peer reviewed and unpublished. So here you have the hypocrisy and farce of the foot and mouth disease epidemic in a nutshell. A test that could save thousands of animal lives was not used, yet a model that advocated slaughter on an unprecedented scale was used.

Modelling your way into a mess

At the outset of the epidemic, Donaldson and his team from Pirbright, as world leaders in foot and mouth disease, were advising MAFF on how to tackle the outbreak. MAFF was the ministry in charge. But all this was about to change. A chain of events took place that shows once again how a small powerful elite runs the UK's scientific decision making even though the consequences of their actions may be catastrophic.

This elite group of scientists, normally members of The Royal Society, have a stranglehold on scientific expression in the UK. Much of scientific government policy on food and related issues is run by scientists based or linked to the 'closed shop' of Oxford and Cambridge. There is a disturbing pattern of many of the influential characters who sit on government committees that have looked into food, farming and GMOs, being both members of The Royal Society and coming from the Departments of Zoology at primarily Oxford, but also Cambridge. As one leading scientist described these, they are the 'closed shops of closed shops'; they are the inner circle of the old boys club.

Three days before Roger Breeze sent his email to Pirbright, Sir John Krebs, the head of the FSA and a Royal Society fellow, convened a secret meeting. Krebs had been a surprise appointment to head the FSA. Sir John, who still works part-time at the Department of Zoology at Oxford, had no official remit to get involved in foot and mouth disease, but asked his old friend and ex-colleague from the department, Professor Roy Anderson an epidemiologist, to undertake some modelling on foot and mouth disease.[31]

Anderson and Krebs have a history of collaboration. In 1996 the government of the day commissioned an independent scientific review of future policy options concerning the control of bovine tuberculosis. The independent review team was chaired by Professor John Krebs and included both Anderson, and also Dr Christl Donnelly, from his team first at Oxford and then at Imperial College. Anderson was a co-author with Krebs of a controversial review of bovine TB known as the 'Krebs Report'.[32]

Anderson is a controversial figure, too. Many scientists feel that he was being groomed for the job of chief government scientist. But in

January 1999 he was suspended from his Chair at the Department of Zoology after his colleagues unanimously passed a vote of no confidence in him. This followed complaints by a female colleague, Dr Sunetra Gupta, whom Professor Anderson had falsely accused of sleeping with a colleague in order to gain a post at Oxford. Gupta demanded, and received, a full apology from Anderson. But Anderson was also criticized for not declaring that he was a trustee of the Wellcome Trust that had in turn been funding his department. In May 2000 Anderson resigned and took his research group to Imperial College.[33]

Anderson's speciality is modelling human disease – such as AIDS, malaria and TB. Due to its human connotations, Anderson has also modelled BSE and its human equivalent vCJD as well as bovine TB. Anderson sits on SEAC, and it is his models that provide much of the guidance on BSE and vCJD forecasting. In addition, in 1999 a member of Anderson's group, Dr Christl Donnelly, had undertaken retrospective analysis of the 1967–1968 foot and mouth disease outbreak, examining how it might have been reduced in scale.[34]

What Krebs wanted was for Anderson to be put in charge of the modelling of foot and mouth disease and the secret meeting on 6 March – to which MAFF was not invited – was the first move whereby those who knew little about an issue – foot and mouth disease – once again sidelined those who did.[35]

On 21 March another meeting took place with Anderson, but this time it was attended by the government's new chief scientist, David King, who had replaced Sir Robert May, who had left to take up the position of head of The Royal Society. May had suggested King as his replacement and had co-written two books with Anderson.

King was about to take over control for handling of the disease from MAFF, and was convinced by the presentation from Anderson's team, more so than that of a team of rival epidemiologists. So King, whose speciality was 'surface chemistry', was about to use Anderson's model to drive government policy. The only problem was that the men now running the UK's response to foot and mouth disease had little background in veterinary science.[36] Although King was to set up a Science Group on which scientists from Pirbright would sit, it was Anderson and his model that won through.

On the night of 21 March Anderson appeared on the BBC's *Newsnight* programme. In a public rebuke to MAFF, Anderson said the 'epidemic is not under control at the current point in time'. He also reinforced the newly introduced 'contiguous cull' that had been introduced for sheep in Cumbria three days before, where every animal was to be killed in a 3km range of an infected farm. But what he was proposing was that the government needed to bring forward the 'preventative' contiguous cull. 'If this cull is applied vigorously and effectively enough you could turn the epidemic in to a decaying process hopefully within you know, a month to two months', Anderson told Newsnight.[37] His model would show that the epidemic would be over by 7th June – coincidentally the day of the proposed election.[38]

Three days later on 23 March the contiguous cull was announced and three days after that the epidemic actually peaked. Despite this, the government pushed ahead with Anderson's recommendation and on 29 March the new targets of '24/48' hour culling were announced – 24 hours for infected farms and 48 hours for neighbouring farms.[39]

That same day the government was told that there was a problem underlying Anderson's assertion. The then Agriculture Minister, Nick Brown, was told by the Deputy Director of the Pirbright Institute, Dr Paul Kitching, one of the leading experts on the disease, that it was based on flawed data and assumptions.[40] Speaking a year later, Dr Alex Donaldson confirmed that the 'the novel models that were used in this epidemic were untested and invalidated'.[41] But still the slaughter continued.

Anderson's model had been put forward as the solution to the epidemic by two members of The Royal Society, who are the government's top advisors on food and science, Sir John Krebs and Professor David King. At the height of Pusztai saga (see Chapter 5), two years previously, Professor Anderson, along with 18 other prominent members of The Royal Society, had written an open letter. It warned 'those who start telling the media about alleged scientific results that have not first been thoroughly scrutinised and exposed to the scientific community serve only to mislead, with potentially very damaging consequences.'[42]

The damaging consequences of the contiguous cull were the needless slaughter of some 2 million animals[43] with the resulting trauma suffered by thousands of farmers that could take a generation to heal. Indeed, the senior vet in Dumfries and Galloway was Roger Windsor, who subsequently described Professor Anderson not as a 'Professor of Epidemiology, but the Professor of Extermination'.[44]

Small-holders speak out

One person who has studied Anderson's models is Alan Beat, formerly a Chartered Mechanical Engineer, who now writes regularly for *Country Smallholding* magazine. With his wife Rosie, he lives on an idyllic 16-acre small-holding in north Devon, keeping a range of livestock including pigs, ducks and a home-bred flock of coloured sheep developed over 13 years for their fleece. During the last few years, Alan has restored the farm's mill wheel and their annual Mill Open Day attracts hundreds of people. The couple run educational visits for local school children, too.

Like farmers across the land, the Beats watched the unfolding foot and mouth disease saga with increasing horror. The couple started to read up everything they could find on the disease, including on the internet, but their theoretical interest became personal, when on Good Friday, 13 April, they were phoned by their neighbour to say that his animals had foot and mouth disease. 'It was six o'clock in the morning when the phone rang and it was our neighbour', recalls Rosie. 'He was really upset. We waited for the phone call from MAFF. It was awful.'

Easter, which is normally a time of rebirth and renewal, was going to be a nightmare for the Beats, as it was for many farmers in 2001. That Good Friday 1 million dead livestock lay in the fields waiting to be buried or burnt. Due to the contiguous cull, the Beat's livestock would have to be destroyed. Later that day the phone call from MAFF came. The Beats responded by telling MAFF that they could only blood test their animals, not cull them; only if they had the disease would they be culled.

The Beats refused to leave the farm for three weeks, barricading the gate. 'At the time we were traumatised and fearful,' says Alan. 'I was having nightmares', adds Rosie. 'I was in floods of tears all the time. Every time the phone rang, my innards would turn to jelly. The whole thing has made us different people, I used to be so trusting.'

'I told them straight that they needed a court order if they wanted to come in here and they never did. Now that tells you quite a lot', continues Alan. 'If they wanted to come in they would have gone for that court order. But they knew that they did not have the grounds to get us. It tells me that the contiguous cull was illegal.'

Their flock was monitored every few days, but in the end the Beats managed to save their prized livestock and now feel vindicated by the actions they took. 'We felt there were grounds on which to refuse', adds Alan. 'In Devon, there were roughly 200 farms that resisted the cull and not one of them subsequently developed the disease.'

Their traumatic experience has propelled Alan and Rosie to discover the truth about foot and mouth disease. 'We came into foot and mouth with no preconceptions', says Alan. 'I am now looking back at BSE and seeing the parallels and looking forward to scrapie and seeing the parallels.' Part of that truth lies in unravelling the foot and mouth models.

'I have looked into these computer models and I am convinced they were wrong at the outset, and that the errors have been covered up', says Alan over a cup of tea in his 300-year old Devon cottage. 'They have been used for a justification of a policy that they wanted to pursue, rather than a scientific examination of the facts about foot and mouth and the way it spread and applying those facts logically.'

Although Anderson's model effectively drove government policy on the disease it was not until May that it appeared in *Science*, a peer-reviewed journal. When Alan Beat studied it, he was horrified. 'You go through Anderson's paper, and it's a very simplistic model. We subsequently learned that it was based on the spread of human sexually transmitted diseases. He said that in public. He took the model and adapted it for foot and mouth. But there are glaring differences between human sexual behaviour and farm animals.'

Alan Beat is not impressed at the lack of common sense and logic shown by the models and the reasoning behind the contiguous cull. The full data have never been published, argues Beat, all we have ever seen are their interpretation of the data. 'But there were more flaws', argues Alan. 'Instead of sheep, cattle and pigs, he assumes a standard hypothetical species of livestock. But there are huge differences in the potential infectiousness and spread from different species, but he did not take any account of that.' In fact cattle are 15 times more susceptible to the disease than sheep.

Another thing that was wrong was the issue of farm infectivity. Any viral disease spreads through the herd rapidly to a peak and then dies away, argues Alan. But Anderson's team assumed it was constant from the moment the farm is infected to the time of slaughter. So Anderson concluded that farm infectivity was equal to one. 'The point is that farm infectivity will be greater than one' retorts Alan and 'when you put farm infectivity being greater than one into his model, it shows that the contiguous cull was unnecessary'.

But Anderson's team did not model this scenario, because, they wrote in their paper, the 'data do not exist'.[45] However the data did exist, argues Beat and because the papers containing this data are co-authored by Paul Kitching from Pirbright 'it should not have been unknown to Anderson's team'.[46] For whatever reason Anderson's team ignored it.[47]

In summary, the model was 'fundamentally flawed through the use of wrong input data and wrong assumptions', argues Alan. 'The original field data from MAFF were wrong; assumptions about airborne spread were wrong; crucial differences between livestock species were ignored; the significance of long-distance spread via movements of animals, personnel and vehicles was underestimated and assumptions of the change in farm infectivity over time were wrong.'

Five months later, Anderson's team published a second paper in *Nature*, and Beat believes this proves that the first paper was flawed and that the contiguous cull was unnecessary. Back in May, the team's first paper declared 'that farms closest to index cases of foot and mouth are at greatest risk of infection'.[48] 'This, of course,' argues Beat, 'promoted the concept of contiguous culling to remove those farms at highest risk before they, in turn, became infectious to others'.

But, in October, the second paper contradicts this by stating 'that most transmission probably occurred through the movement of animals, personnel or vehicles, rather than through animal contact or windborne spread'. The October paper also admitted that animal restrictions and tight biosecurity were actually more important in stopping the disease than culling.[49] 'The admission that the original data were completely wrong' says Beat, 'entirely destroys the rationale for contiguous culling.'[50]

Not only did the Beats believe that the models were flawed, but that they had no bearing on reality. 'We were sitting here in Devon', says Alan Beat. 'We were watching the disease apparently jump four, five, six miles at a time. It would be four miles away; it would be eight miles away. There was no infection to neighbouring farms. Or it certainly was not as common place as the computer modellers were trying to tell us. We were watching it on the ground and thinking it does not make sense. This does not follow the predictions of the computer model.'

The final flaw in the contiguous cull is that it probably had no bearing on the epidemic anyway. 'If you start killing contiguous farms on the 29th March, it will be the 9th April before you make any impact whatsoever because of the incubation time, and by that time it was all over,' argues Beat. 'If the culling policy had an impact you would see a dramatic reduction after the 9th April, and it didn't happen, it just carried on downwards, all we did was pile up the dead bodies. We had nearly 100,000 carcasses awaiting disposal in Devon at one point. That is all it did.'

Despite this, civil servants continued to say that the cull was a necessity. Ben Bennet, MAFF's divisional veterinary manager said the cull had been 'very effective in reducing new infections.'[51] The cull was, to all intents and purposes, illegal too. Under the current Animal Health Act of 1981, the law under which the government justified the cull, only the culling of infected stock or stock exposed to infection was legal. Therefore the government 'acted criminally in coercing or forcing farmers to submit to such culls'.[52] It was also the reason that MAFF backed down when challenged by the Beats and other Devon farmers.

The government scientists at Pirbright agree that the models were wrong too. Dr Alex Donaldson says that the policies – the contiguous cull and 24/48 slaughter policy proposed by Anderson were 'extreme' and 'diverted scarce resources'.[53] Further research by Donaldson and his colleagues showed that the contiguous cull was flawed and that for airborne cases of the disease the virus was likely to be carried just 100–200 metres, not the 3km used in the contiguous cull.[54] Dr Paul Kitching, from Pirbright one of the leading experts on foot and mouth disease, also described the contiguous cull policy as fundamentally flawed.[55]

In July, Kitching gave evidence to the EU Inquiry. He concluded that the 'predictive models produced were deficient in a number of input parameters'. The PanAsia strain causing the outbreak did not spread significantly as an aerosol, said Kitching and because the models did not accommodate the delayed diagnosis of foot and mouth disease, their predictions of the rate of spread to new premises was inaccurate.[56]

Kitching also laid bare the misinformation from Ministers that the Science Committee had been in favour of the cull. 'On numerous occasions during meetings of the Science Committee, both myself and Dr Alex Donaldson expressed concern about the validity of the policy derived from the models', said Kitching.[57] This is in direct contrast to what the Agriculture Minister, Nick Brown said in May. 'I know it is incredibly distressing when farmers are told that their apparently healthy animals have to be culled, either as dangerous contacts or as part of the contiguous cull. But the advice both from the veterinary authorities and from the Scientific Steering Group to the government was very clear indeed.'[58]

It was clear, but the advice from the experts was against Anderson's model and the cull. Kitching, who was once described as a 'gem of veterinary science,' was so disillusioned by the foot and mouth disease that he left Pirbright and took up a new position in May 2001 in Canada.[59] On 1 May 'I asked for a summary of results generated at Pirbright', Kitching told the EU, of 1876 premises that had been slaughtered, samples from 52.76 per cent were negative on laboratory tests. This was reported to the Science Committee on 2 May.[60] Other data have now confirmed that of a sample of 720,000 culled animals, only 327 showed signs of the disease, a percentage of 0.045 per cent.[61]

'In my opinion' Kitching concluded in his evidence, 'the takeover of the programme from MAFF by the Science Committee, which was heavily influenced by modellers with very limited practical experience of foot and mouth disease, resulted in the unnecessary slaughter of possibly as many as 2 million animals.[62] The EU Parliament's Inquiry into foot and mouth disease was equally damning saying that the government's decision to slaughter over 6 million animals did little or nothing to stop the disease. It also called on vaccination to be used as the tool of first resort.[63]

To vaccinate or not to vaccinate

During the epidemic, the Beats, like many farmers, could not understand why vaccination was not being used. 'I found it deeply offensive that here we had an animal disease for which an effective vaccine was available, that could legally be used, but we were not allowed to use it', says Alan. 'Eleven million animals were slaughtered, often in conditions of extreme cruelty, especially these so-called "welfare culls". Why not use the available scientific knowledge that we had to solve this problem in a way that is animal friendly and kinder to humans and the environment?'

Many of the world's experts, such as Fred Brown and Dr Keith Sumption, argued for vaccination along with thousands of small farmers and the 3 million or so members of environmental and conservation groups, such as the Soil Association, RSPB and Wildlife Trusts.[64] Dr Sumption from Edinburgh University, an international specialist in animal diseases, called vaccination 'a rational response to an exceptionally infective condition'.[65] In March, Sumption pointed out that the mass culling policy had failed and that vaccination could end the epidemic in weeks.[66]

Professor Fred Brown would have vaccinated, too. 'It works, you make the rings – whatever size of ring you want – you come in from the outside and vaccinate inwards from the outside. It has been shown to work in lots of places,' he says, particularly in Holland, which had a foot and mouth disease outbreak at the same time. Dr Simon Barteling, another foot and mouth disease expert would have vaccinated too. He has participated in no less than 23 missions on foot and mouth disease in a multitude of countries. Paul Sutmoller, who since 1985 has worked for the American arm of the World Health Organization, the World Bank, and US Department of Agriculture, also advocated vaccination.[67]

Brown, Barteling and Sutmoller spoke at meetings in Bristol and Cumbria to discuss vaccination. An issue that is governed 'not by science but international trade' argues Brown. 'If a country has vaccinated no other country will import from them, in general, but, in reality, the rules are broken the whole time.'

'The rules state,' adds Alan Beat 'that if you vaccinate and kill you can export in three months, but if you vaccinate to live, you wait 12 months. You had a nine-month penalty. If you slaughter to get rid of the disease and there are no new outbreaks for three months, then you are presumed to be clear of the disease. But if you vaccinate to live then there is the perceived risk that the vaccinated animals could reinfect unaffected stock, if they come into contact with unvaccinated stock. As far as I have been able to trace, there is no recorded case of a vaccinated animal ever infecting another.'

Since the outbreak, the rules have been changed. The OIE has relaxed the 12-month rule back to six months, if you vaccinate to live. But experts like Fred Brown continue to argue that six months should be reduced to three, so that vaccinating to live is the same as slaughtering in terms of trading penalty. 'Up to now it has been dominated by trade and politics', says Beat.

Traditionally, mainland Europe vaccinated, whereas the UK slaughtered. 'Europe banned vaccination in 1992 because of world trade,' says Brown. 'But historically it had been used to treat the disease. It really worked. It was a huge success story. But then it was a case of 'if you vaccinate, you might be harbouring the virus, and therefore we don't want your animals'. It was the UK that persuaded the rest of Europe to stop vaccinating. The UK has never vaccinated. It has always had the advantage of being able to restrict or prevent the movement of animals.

In the 2001 outbreak, Holland originally vaccinated against the disease, but then due to concerns over trade, the animals were killed. 'They slaughtered 260,000 animals in Holland, 200,000 had been vaccinated and didn't need to be killed, but were killed purely due to those restrictions on trade', argues Beat.

One country that suffered a similar outbreak to the UK in 2001 was Uruguay. But only 7000 animals were killed and they vaccinated their cattle herd of more than 10 million animals in three weeks. 'They set about a mass vaccination of the entire cattle population,' says Alan. 'They vaccinated infected premises and did not slaughter afterwards, just quarantined them until the epidemic was over. Sheep were not touched. It started in April and their last outbreak was in August.'

Uruguay started re-exporting to Europe in November. 'They started their outbreak later and eliminated their outbreak much more quickly, with very little loss of life. It cost them $13 million. It tells us that vaccination is a much better alternative', adds Alan; even taking into consideration the fact that Uruguay's main export is beef, which is easier to export after a foot and mouth disease epidemic.

But we have never vaccinated in this country and so strong was the anti-vaccination message that government vets who proposed vaccination were told they could be sacked.[68] The National Farmers' Union (NFU) argued against a policy of vaccination as a 'leap in the dark'. Ben Gill, the head of the NFU, called vaccinated animals, 'the walking dead'.[69] The NFU were against vaccination – due to their large cattle breeders and exporters being against it, not the small farmers. 'Gill told me in person that he took the credit for stopping vaccination', says Alan, who interviewed Gill for a documentary on foot and mouth disease. 'Gill himself says it was the NFU that stopped it.'

The day before the Beat's neighbouring farm had been confirmed as an infected premises, Blair held a secret crisis meeting at Chequers which Gill did not attend, but arch rivals such as the Soil Association did, along with the major food retailers. At the meeting it was announced that vaccination was the way forward. But the following Tuesday, the NFU produced a list of some 51 technical or legal questions that they demanded be answered before vaccination took place.[70]

Alan Beat feels that part of the misinformation that the NFU wanted clearing up came from the government. 'One of the questions had asked whether the government would be the purchaser of last resort, if all else fails', says Beat, 'and the government said no. Well we subsequently found out that it is written into the EU Directive on foot and mouth control that, if vaccination is used, it is perfectly legitimate for farmers to be compensated, if their milk or meat or other produce is financially damaged. The provision is already there within the EU Directive for full compensation.'

The not-so-public, public inquiries

Like others, the NFU asked for a public inquiry, but they were to be disappointed. Instead the government ordered three inquiries, none of which would be public. 'The main issues that we really wanted to be sorted have slipped down between the chinks of the three,' says Rosie Beat.

It is easy to see why a public inquiry was resisted: once again the focus was not to blame Ministers or individuals, just as the Phillips Inquiry had been on BSE. 'The nation will not be best served by seeking to blame individuals', wrote Dr Iain Anderson, in the introduction to the 'Lesson Learned' Inquiry.[71] This angers the Beats. 'That is what accountability means, isn't it, and that is what public office means, that you are responsible for your actions,' says Alan Beat.

So it was politically acceptable for Labour to order a public inquiry into BSE – an £8 billion disaster caused by the Conservatives – but it was not acceptable to find out what happened with Labour's own £8 billion agricultural mess of foot and mouth disease. The Royal Society was asked to examine science and infectious diseases, Dr Iain Anderson was invited to examine 'Lessons Learned' and Sir Donald Curry chaired the Policy Commission on the Future of Farming and Food.

The chair of The Royal Society working group was Sir Brian Follett, who as a Visiting Professor at the Zoology Department at Oxford is yet another academic with links to the department asked to Chair an inquiry into a food or farming issue. Included in his team was Dr Angela Maclean, another member of the Zoology Department at Oxford and Professor Patrick Bateson, The Royal Society's Biological Secretary and from the Zoology Department at Cambridge. Sir John Krebs' deputy at the FSA at that time, Suzi Leather, was also included.

'What made this inquiry so open to suspicion', wrote Christopher Booker and Dr Richard North in the *Private Eye* special edition on foot and mouth, 'was the network of contacts intimately linking it to the little group of scientists who had been at the centre of the crisis since March, and who had been personally responsible for some of the most questionable aspects of the way it was handled.'[72]

There were two surprise members, Professor Fred Brown and Jeanette Longfield, from Sustain, the Alliance for Better Food and Farming. Ill-health kept Brown away from most of the meetings. 'It was intended to say, what do we do next time, not what should have been done last time', he says. 'The report was pretty fair.' It argued that 'emergency vaccination' should be considered from the start of any outbreak, with the long-term solution to develop a vaccine against foot and mouth, although 'extensive culling' may still be necessary.

Meanwhile the Chair of the Future of Farming report was Donald Curry, who had been in charge of the Meat and Livestock Commission (MLC) at the time of the BSE Inquiry. The MLC had been criticized by the Phillips Inquiry for failing in its statutory objectives and highlighted how 'hyperbole displaced accuracy' in its campaigns.[73]

Iain Anderson was an old ally of Margaret Beckett, the new Minister in charge at Defra and who had served as Blair's special advisor on the Millennium bug.[74] Brown felt that Anderson's Lessons Learned report was 'a bit of a white wash'.

But probably the most amazing fact about the two foot and mouth-specific inquiries – Anderson's and The Royal Society's that recommended that the government should do things differently, is that they probably made no difference at all.

When interviewed by Alan Beat, the Chief Government scientist David King said that, if there was another outbreak, the government would do the same again. 'So we have The Royal Society report that recommends vaccination as a tool of first resort', says Alan Beat. 'We have a Lessons Learned Inquiry that says that vaccination must be considered as part of the armoury. But the Chief Scientist, David King, says "no we carry on slaughtering, just like last time because the new tests have not been validated".'

Beat says that two years later they have made no attempt to validate them. 'Nothing has been learnt; nothing is going to change. I found that the most chilling answer. He can't change the policy, because to do so would be to admit they were wrong.'

Not only that but things could get worse. 'They are still looking to push through the Animal Health Bill. They have gone to more extreme violence against people and animals.' It had been announced that in the autumn of 2001 the Animal Health Bill would be amended to make it legal for Defra to enter any premises at any time and kill any livestock, with the owner allowed no right to appeal. Farmers, Opposition MPs and vets opposed the Bill. 'The slaughter being extended to all animals on premises designated by a computer, because they are deemed 'contiguous' gives considerable problems to veterinary surgeons', responded Roger Green, the President of the Royal College of Veterinary Surgeons, 'as they have been and could be instructed to kill animals without any good scientific reasons'.[75]

In November 2002, Margaret Beckett, on behalf of Defra, responded to the foot and mouth disease inquiries. 'The basic strategy,' she said 'is that, as a first step, animals infected with foot and mouth disease and animals which have had contact with them, have to be culled'. But Beckett conceded that 'additional action' could include 'emergency vaccination to live'.[76] Alan Beat remains to be convinced: 'I don't think much has changed', he says 'apart from a lot of political hot air. The reality remains that they will start with slaughter and only vaccinate if they are forced into it'.

At the end of 2002, David King, the government's chief scientific advisor, was knighted in the New Year's Honours list, partly as a reward for his handling of the foot and mouth disease crisis.

Waste not, want not

Not everyone suffered during the foot and mouth disease crisis. The Scottish firm, Snowie, made £38.4 million from disposing of carcasses from the north of England and Scotland and from the construction and management of the mass burial site at Great Orton.[77]

The Snowie brothers are reported to be enjoying their new-found wealth, recently buying their own yacht firm, and expanding their property portfolio to include castles and mansions.[78] The row over the payments to Snowie was further ignited by the revelations that Euen Snowie, one of the four brothers who run the Snowie companies, gave the Scottish Labour Party £5000 in June 2001, just after the height of the foot and mouth disease crisis. Although the Commons Public Accounts Committee and the National Audit Office were called to examine the donation, Snowie denied any wrongdoing.[79]

Before the foot and mouth disease crisis, Snowie had been growing rapidly as a waste contractor dealing in part of the BSE problem that no one wants to know about and also with sewage sludge, a burgeoning waste problem since dumping at sea was stopped at the end of 1998.

The waste disposal industry is a controversial one. The industries argue that they are doing an incredibly beneficial worthwhile job. Their critics point out that it is precisely because it is so crucial – that they get rid of a problem that no one wants – that the regulators give them an easy time. Much waste application is actually 'exempt' from licensing. 'They have the government over a barrel, as far as I am concerned', says one environmental health officer.

The small communities of Blairingone and Saline, which lie under the beautiful Ochill hills north of Edinburgh, were shocked to see Snowie get the foot and mouth disease contracts, and wondered why such a controversial company should have been given the contract. The land surrounding the village had been an open cast mine, but the fields were bought by Snowie in February 1997. Since then, the village has had a 'raging war' with Snowie, according to local MP and Deputy Presiding Officer of the Scottish Parliament, George Reid. Complaints about noxious odours emanating from Snowie's site commenced a month after the waste contractor started operations and have continued ever since.[80]

Due to their growing concerns, the community formed the Blairingone and Saline Action Group or BSAG. 'They started bringing sewage sludge and other wastes, causing a huge stench in the area, that's what people noticed first, before they realised what it was,' says Duncan Hope, one of BSAG's co-founders. 'We formed BSAG as the local council had done nothing about it. Because it wasn't just happening here, it was a national issue, we felt the best course of action was to get new legislation or to tighten up the current legislation.'

Under current legislation, Snowie is allowed to spread some 250 tonnes per hectare of 'excempt wastes' such as human sewage sludge and cake tannery, fishery, distillery and food processing effluent, paper crumble, and abattoir waste such as blood and guts onto land.[81]

Although the practice is largely unmonitored, on three occasions Snowie has been fined for breaching environmental standards. Snowie Ltd was fined £3000 in February 2000, £1000 in October 2000 and £5000 in June 2001 for disposing of waste in a manner likely to cause pollution of the environment and/or harm to health.[82]

Nor is Snowie alone in spreading blood and guts. In 2002, another Scottish waste contractor, Beneagles, had also generated complaints from locals after spreading blood and guts from abattoirs. George Reid MSP called for a ban on the practice in the summer of 2002,[83] but at the time it was legal, although authorities knew it posed a risk to health.

The relevant regulatory authority in Scotland is SEPA – or the Scottish Environmental Protection Agency. Following widespread public concerns 'about potential health effects and malodour nuisance created by spreading organic wastes to land', SEPA commissioned a report in January 1998, called the Strategic Review of Organic Waste Spread On Land, known as the OWL report. Its two primary conclusions were that 'the current approach to the regulation and management of organic waste spread on land is inadequate and inconsistent, leading to practices which pose a risk to the environment and pose potential public, animal and plant risks'. Because of this there was 'a lack of public confidence in the practice of spreading organic waste on land'.[84]

One of the key recommendations was that 'blood and gut contents should be prohibited from being spread on land'. The report by SEPA was sat on for two years by the Scottish Executive, an act labelled 'outrageous' by Bruce Crawford MSP, Shadow Minister for Transport and Environment.

SEPA has been dismissed by BSAG as a 'toothless, useless organisation'.[85] In Scotland, there is no central register for wastes such as blood and guts from abattoirs, as these are considered legally 'exempt' from licensing. In 1997, the Agency estimated some 26,000 tonnes were put on the land in Scotland.[86] The practice remained widespread although it is now due to be banned under EU law in 2003. Snowie has admitted spreading some 675 tonnes of blood and guts onto one farm in Blairingone, near Dollar in 1999–2000.[87] But the company denies that it now spreads bovine blood to 'agricultural land'.

BSAG was aided in its campaign by the new Scottish Parliament and George Reid. Through Reid, the Blairingone community filed a petition PE 327 at the Scottish Parliament on 12 December 2000. It asked the Parliament 'to request that legislation be revised to ensure that public health and the environment are not at risk from the current practice of spreading sewage sludge and other non-agriculturally derived waste on land in Scotland'.[88]

The MSP chosen to investigate the community's petition was Andy Kerr MSP. 'I learned that I could go to a second-hand shop and buy a tractor, some injecting equipment and an articulated container, then go to an abattoir to collect blood and other products', Kerr told colleagues. 'Without notifying the neighbours, I could get a piece of land, go out at 2 o'clock in the morning and spread the products on the land. In these days of E. coli, BSE, pathogens and other such critical issues, I find it surprising that we do not have greater control over the disposal of organic waste.'[89]

The Scottish Parliament's Transport and Environment Committee published Kerr's report in March 2002. In an overwhelming victory for the Blairingone and Saline residents, the Committee expressed 'surprise and concern' that the law permitted individuals to spread waste products such as blood and guts from animals and sewerage sludge 'without proper regard for health and environmental factors'. Nor did it allow residents a proper mechanism to substantiate complaints against odour problems.[90]

'We still have not had an investigation into the illnesses', says Duncan Hope, the Chair of the BSAG. The community, who stress they have no proof that the waste operations are causing illness are still demanding a proper investigation. Duncan Hope and BSAG's petition is now held up as a showpiece of how community action can work in the new political landscape.

For all their hard work, though, the community has not yet managed to change the legislation. 'They are advocating the use of wastes on land, they don't know what is in these wastes, and what damage they could do or are likely to do', says Hope. Snowie dismiss the allegations against them and the community's petition, arguing that BSAG was 'not representative of the community' and that it had 'instigated a campaign of misinformation against the Snowie Group of Companies'. The company also pointed out that 'Snowie Limited does not recycle bovine blood to agricultural land. Indeed Snowie Limited has urged the regulatory authorities to stop this practice'.[91]

Snowie's response angered George Reid MSP. 'I deeply regret that Snowie Ltd chose to denigrate them [BSAG] in its evidence. Among the signatories to the petition are more than 300 people who are resident in the immediate locality of Blairingone and Saline. That is a much higher participation rate than was achieved in the general election.'[92]

Blood and guts from cattle were legally spread on land until 2003, when it should stop under incoming EU legislation. The reason it was stopped, says the Environmental Agency, was concern within Europe of blood and guts spread on land 'being a disease issue, which is clearly linked to TSEs'.

But we have known for 20 years that slaughterhouse waste could be a problem. There had been concern within MAFF about the practice since the late 1970s and it has been known since the late 1980s that practices were inadequate.[93] So it has taken over a decade for this potential infection route to be closed, as once again precaution was thrown to the wind.

It is here that we go full circle and return to the niggling uncertainties about BSE. It will be years before we know whether there is long-term groundwater contamination from BSE-infected cattle being buried during the foot and mouth disease crisis. We do not know how much infected BSE blood is lying dormant in the soil. It looks like whatever risk there is will be small. But other transmission routes remain and as recently as February 2002, SEAC noted that 'many scientific uncertainties relating to the propagation and degradation of BSE infectivity in the environment' remained.[94]

But whilst we still do not know the long-term health impact of the BSE epidemic or indeed when that epidemic may be over, we stand on the threshold of a widespread adoption of new agricultural practices such as genetic engineering. Like BSE there are many unanswered questions remaining, and also like BSE and foot and mouth disease, the scientists who have spoken about the dangers have been publicly vilified or marginalized.

CHAPTER 5

Hot Potato

'The public has the right to expect the very highest standards of food safety. Confidence in the safety of the food we eat has been severely undermined in recent years and I am determined to rebuild that trust'

Tony Blair[1]

'They picked the wrong guy – I will kick the bucket before I give up'

Dr Arpad Pusztai

As we witness the dawn of the biotech revolution, Dr Arpad Pusztai is a scientist who is convinced that he has uncovered vital evidence that shows there are potential major health risks with GM crops. Pusztai was catapulted from an unknown laboratory scientist based at the Rowett Research Institute in Aberdeen to the forefront of a raging debate about the safety of GM foods, when he spoke on the *World in Action* TV programme in 1998.

Overnight the Hungarian-born scientist, with some 35 years lab experience, found himself at the centre of an international media spotlight. The controversy would put him on a collision course with the UK and US governments, the biotech industry and the scientific establishment. His 150-second interview lead to Pusztai being suspended, silenced and threatened with losing his pension. His wife, Susan Bardocz, who also worked at the Rowett for 13 years, was eventually suspended too. Their research was locked up. Scientists and politicians alike vilified Pusztai.

The fall-out left many unanswered questions, and just like BSE, the more you look, the more uncertain the existing answers become. Although the Pusztai story became well-known within the GM debate, the full story of what happened has never fully been told and remains a controversy.

As we search for answers as to whether GM foods are safe, two questions stand out. Given such a huge controversy over Pusztai's experiments, and the preliminary nature of their findings, why were the political and scientific establishments so intent on rebutting him? More importantly why have the experiments never been repeated? In the answer to those questions lies the kernel of truth. But the truth will only unfold after the UK's Prime Minister, Tony Blair, clarifies whether he had any role in Dr Pusztai's sacking.

The issue has had very personal consequences. Pusztai has suffered two heart attacks and the saga has left him and his wife, Susan, needing permanent medication for high blood pressure. Pusztai is still angry about the whole affair. His only crime was to speak out, in his words, according to his conscience: 'I obviously spoke out at a very sensitive time. But things were coming to a head with the GM debate and I just lit the fuse', he says. 'I grew up under the Nazis and the Communists and I understand that people are frightened and not willing to jeopardise their future, but they just sold me down the river.'

Dr Arpad Pusztai

Pusztai's story begins in post-war communist Hungary. After the Hungarian revolution was crushed by the communists, the young Pusztai, a chemistry graduate, escaped to refugee camps in Austria and from there to England. By 1963, having finished his doctorate in biochemistry and post-doctorate at the Lister Institute, he was invited to join the prestigious Protein Chemistry Department at the Rowett Research Institute, then headed by Dr Synge, a Nobel Laureate, and the co-discoverer of chromatography.[2]

The Rowett Institute was founded in 1913 to conduct research into animal nutrition, and has broadened its aims to 'define how nutrition can prevent disease, improve human and animal health and enhance the quality of food production in agriculture'. It has become the pre-eminent nutritional centre in Europe.

Dr Pusztai was put to work on lectins, plant proteins that were going to be central in the GM controversy years later. Within a year Pusztai had won acclaim for the Rowett by discovering glyco-proteins in plants, which until then had been identified only in mammals. Over the intervening years, Pusztai became the world's leading expert on plant lectins, publishing over 270 scientific studies, and three books on the subject. Two books were co-written with his wife, Susan.

Pusztai became one of the Rowett's most senior and renowned scientists. At the age of 60 Pusztai should have retired, but he was asked to stay on by the Director, Philip James. Pusztai won a prestigious Leverhulme Fellowship and became a Fellow of The Royal Society in Edinburgh.

In 1995, the Scottish Office Agriculture Environment and Fisheries Department commissioned a three-year multi-centre research programme under the coordinatorship of Dr Pusztai into the safety of GM food. The other participating institutions were the Scottish Crop Research Institute (SCRI) and Durham University Biology Department. At the time there was not a single publication in a peer-reviewed journal on the safety of GM food.[3]

The fact that the Scottish Office was primarily in control, rather than the Rowett later added to the controversy. 'These experiments were organised without any reference to myself or the Directorial system at the Rowett', maintains Professor Philip James, head of the Rowett at the time, who says the experiments were going on 'completely outside his control'.

'The background behind the GMO work was to try and balance the fact that a lot of the work on biotechnology was happening with very little regard to environmental or human nutritional consequences', says a person who was involved in the initial negotiations, but who has since left the Scottish Office. 'Dr Pusztai was a renowned scientist on the international scene. He had an international reputation. The fact that Pusztai was kept on at the Rowett is very unusual. Anybody who stays beyond 60 does so for quite exceptional and outstanding reasons. The ability to deliver on results is part of the assessment process we went through.'[4]

The scientists' primary task was to establish credible methods for the identification of possible human/animal health and environmental hazards of GM. The idea was that the methodologies that they tested would be used by the regulatory authorities in later risk assessments of GM crops.[5] For the first time, independent studies would be undertaken to examine whether feeding GM potatoes to rats caused any harmful effects on their health, bodies or metabolism.[6]

All three institutes had links to the biotech industry. Also connected to the project were the growers of the potato, Axis Genetics, later to become Pestax Limited. Scientists paid by the company were working at the University of Durham. Axis Genetics and the Rowett had a profit-sharing agreement should the potatoes ever be commercially grown. 'Nothing is ever done out of the goodness of our hearts', says Pusztai. 'In those days we thought everything was going to be all right and here we are doing work that would make the potato commercially valuable. We do work at the expense of the taxpayer. But if a private company was going to make money out of it, it is fair enough to share that profit. That was why we had a profit-sharing agreement.'

Although Pusztai's experiments were later criticized for having 'flawed' methodology, his team fought off competition from some 28 other research organizations from across Europe to be awarded the £1.6 million contract. The project methodology was also reviewed and passed by the Biotechnology and Biological Sciences Research Council (BBSRC), maintains Pusztai. The BBSRC is the leading funding agency for academic research and training in the biosciences at universities and institutes throughout the UK. 'What speaks volumes is that the grant was awarded to us and not to the other 28 competing research organisations. They don't give that sort of money without checking it out. It is standard practice and makes common sense,' argues Pusztai.

The theory behind the modification of the potatoes was simple. For years Dr Pusztai had explored the beneficial effects of lectins in foods as well as in nutritional supplements and pharmaceutical agents. Lectins can affect the digestive systems of insects and can act as natural insecticides. Arpad's work had shown that one such lectin called GNA (*Galanthus nivalis*), isolated from the snowdrop, acted in this way.[7] Pusztai had worked on the snowdrop lectin since the late 1980s. 'The GNA gene was selected because our studies had shown that GNA caused no harmful effects on the mammalian gastrointestinal tract', says Pusztai.

The thinking was that, if you could genetically modify a potato with the lectin gene inside it, the potato could have an inherent built-in defence mechanism that would act as a natural insecticide, preventing aphid attack. Because it looked promising, the snowdrop gene had already been incorporated into several experimental crops, including rice, cabbages and oil-seed rape.[8]

But lectins can also affect the mammalian gut. Pusztai had already shown that the lectin from the red kidney bean could 'induce marked intestinal damage to the mammalian gut', which is why we soak and boil red kidney beans. So one of his lingering concerns was whether lectins introduced in genetic engineering could damage the mammalian gut as well as pests.[9]

James says that Pusztai did a 'great service' with lectins. 'He highlighted the fact, in talking to the plant biotechnologists, that they were pretty clueless about the potential biological effects of lectins. They just thought they were wonderful. They had never perceived that if they knocked the hell out of insects' guts, they might knock the hell out of human guts too'.

After the three research institutions divided up the tasks ahead, the scientists at Durham University and the SCRI were to look at the effects of GM plants on certain target pests, and beneficial insects. The Rowett scientists' main task was to see if potatoes genetically modified with the GNA lectin were 'substantially equivalent' to parent potatoes and to undertake the rat-feeding trials. Dr Pusztai and his team were not expecting any trouble.

But by late 1997, the first storm clouds were brewing at the Rowett. Preliminary results from the rat-feeding experiments were showing totally unexpected and worrying changes in the size and weight of the rat's body organs. Liver and heart sizes were getting smaller, and so was the brain. There were also indications that the rats' immune systems were weakening.[10]

The results were causing disagreements between the Rowett and the other project members from Durham and the SCRI, who were against disclosure of the results. 'We almost had fisticuffs, almost violent discussions at these meetings', recalls Pusztai. 'They tried to question our competence. I demanded to sort it all out.' To resolve the issue, Pusztai argued, the Scottish Office examined his methodology. 'It was clear that we had professional competence', says Pusztai. A person close to the project from the Scottish Office says that: 'The projects were all subject to scientific review, they were reviewed inside out by external and independent referees'. But the relationship between Dr Pusztai's team and the other research institutes continued to deteriorate.

Pusztai carried on with his experiments, working with a close colleague, Stanley Ewen, who is the one of the top pathologists in Aberdeen, having 37 years experience. Ewen has never had any contractual arrangements with the Rowett, as he is employed by the University of Aberdeen. His academic involvement with Pusztai, he says was for the 'love of science'.

Like Pusztai, Ewen has reached the top of his field. Over his career he has published 160 scientific papers, which he feels was 'not bad'. He is also a deeply Christian man, who plays the organ at his local church, and is a passionate fly-fisher, especially on the upper Dee, one of Scotland's most famous rivers.

Ewen and Pusztai have had a long working relationship, publishing their first paper in the mid-1980s. Since then they have spent a decade working together, including presenting papers at EU meetings. The pathologist maintains the work he has done with Pusztai is some of his best, because 'each time the experiments seem to get better, such as the quality of the tissue samples, because we were getting so experienced at it. In my judgement it was the highest quality work'.

But by the beginning of 1998, scientists at the Rowett, including Pusztai and his boss Philip James, were becoming concerned about GM developments and the inadequacy of trials on GM foods, specifically GM maize. James, who was a member of the Government's Advisory Committee on Novel Foods and Processes (ACNFP) asked Pusztai to assess the validity of licensing applications from Ciba-Geigy and

Monsanto, amongst others (see Chapter 7). Pusztai's reply was that the tests were inadequate.[11]

Professor James has headed the Rowett since 1982, and is one of the world's leading nutritionists. In the mid-1990s, not only was James on the ACNFP, but he also had nearly 20 other professional work commitments on other advisory panels, bodies, committees and editorial boards.[12] Even he was not immune to external pressure and says that he was 'not re-appointed' to one committee when it was realized he was writing a paper that highlighted the problems of GMOs.[13]

James' primary responsibility, however, was as the Director of the Rowett, although his work commitments often kept him away from Aberdeen. By the late 1980s James had fallen out with the Tories, but became friendly with the then leader of the opposition, Tony Blair. The two men were said to get on. It was no surprise that James was chosen by Blair to set up the blueprint for a Food Standards Agency (FSA) whilst Labour was still in opposition. The James report, as it was known, was published a week after Labour came to power in May 1997.[14]

James was in the running to lead the FSA, but the political storm that surrounded Dr Pusztai's TV appearance on *World in Action* put paid to his chances and Sir John Krebs landed the coveted job. James left the Rowett in 1999 and is currently chair of the International Obesity Taskforce. James was later to tell Pusztai 'you destroyed me'. 'I was blamed for Arpad's behaviour,' he says. 'I got the blame when I didn't even control the damned experiments. I think I have been absolutely honourable throughout.'

Adding to James' and Pusztai's anxiety were the results beginning to appear from Pusztai's lab. Early in 1998, with the approval of Professor James, Dr Pusztai appeared on *Newsnight*. 'We are putting new things into food which have not been eaten before', Pusztai said on the programme. 'The effects on the immune system are not easily predictable and I challenge anyone who will say the effects are predictable.' 'Nobody phoned me afterwards,' recalls the lectin specialist. 'I was happy with what he said on *Newsnight*', says James.

The following month, *BBC Frontline Scotland* interviewed James. 'Once the BSE problem is solved, if it is solved', argued the Professor, 'then I think the big public concern is going to be about the huge array of genetic manipulations that we are going to be seeing in the food chain. I think the perception that everything is totally straightforward and safe is utterly naïve.' James also said he was struggling with the challenge that 'the power of the major companies is huge and the question is how are we going to cope and make sure that things are done appropriately'.[15]

But Pusztai's *Newsnight* appearance had caught the eye of Granada's *World in Action*, who approached the Rowett for an interview with him in

April. 'It was common knowledge in the country in scientific circles that there were no other groups doing this type of research and therefore it was quite natural that they came to us', says Dr Pusztai. 'I didn't look them out'.

With the full backing of Professor James, filming took place in late June. Dr Pusztai recalls the Rowett press officer being present for most of the time. He had, he believed, the full backing of his boss and the senior management team of the Institute.

But the *World in Action* programme was experiencing technical difficulties and its transmission was delayed until Monday, 10 August. On the Sunday, 9 August, the media speculation began to increase about the programme when *World in Action* issued a Press release, which declared: 'New Health Fears Over "Frankenstein" Food'. It read 'World in Action reports on the new research from Britain's foremost food science lab, the Rowett Institute in Aberdeen. Scientists there have discovered that rats fed on genetically modified potatoes suffered stunted growth and damage to their immune systems after 100 days'.[16]

The following morning momentum built, with Pusztai appearing on *GMTV* opposite Dan Verakis from Monsanto. 'I think the important thing here is that the safety of the technology at large is not in question,' said Verakis. 'What we are talking about is potentially some issues surrounding one gene, this is one gene out of hundreds and thousands that could be used in biotechnology'.[17]

Two Rowett press releases that day backed Pusztai. The first was from its Chairman, James Provan, MEP calling the work 'of strategic importance to our country and European Union consumers'.[18] The second acknowledged Pusztai as a 'world expert on the impact on dietary lectins. As a result of Dr. Pusztai's work, Professor Philip James, the Director, alerted MAFF several years ago to the possibility of inappropriate genetic material being inserted into foods unless account was taken of its potential biological properties'. It noted that due to this 'MAFF changed its system of scrutiny for novel foods and this also led to changes in the European approach to novel foods.'[19]

The Con A controversy

However, it is at this point that Pusztai and the Rowett begin to argue over the series of events. In disputed circumstances, people started talking about a different – and crucially poisonous – lectin from southern America called Concanavalin A, known for short as Con A. How this happened is still disputed, but the error appeared in the second Rowett press release on the Monday. The crucial difference is that Con A is a known toxic lectin, whereas the snowdrop – GNA – was supposed to be non-toxic.

Pusztai says he never talked to people about Con A. 'I still don't know how the situation came about, this mix-up did not come from me. Some of the information was thrown in as a red herring from someone, but not Professor James'. Pusztai also says that he did not see the second press release that mentioned Con A.

Professor James says he personally took the second press release to show Pusztai, although in parliamentary evidence he later conceded that 'Dr Pusztai did not see that final check'.[20] James says that on Monday afternoon Andy Chesson, his deputy, did three hours of interviews talking about Con A, and it was not until the Tuesday – the day after the programme – that they found it to be untrue that the experiments had been done with Con A.

Pusztai points out that having worked on GNA – the snowdrop lectin – for over a decade, he was very unlikely to start talking about a different one. Stanley Ewen is convinced too that Pusztai did not get muddled. 'Arpad has always had a clear vision. He is certainly never muddled. He was on top of the whole business.'

'I think that is the most likely source of the misinformation, and that the whole Con A potato story originates with Monsanto', speculates Pusztai. Indeed on *GMTV*, earlier that morning Dan Verakis from Monsanto, without mentioning the gene's name, implied that Pusztai's results were 'no great surprise'. 'It is well known that that particular gene that was tested creates a protein that could have some issues with the immune system,' said Verakis, talking about Con A, not the snowdrop lectin.[21]

On *GMTV* Pusztai had not mentioned any specific gene, keeping to his agreement with James, saying that 'we knew in advance that there might be some problems so we used it as a positive control.'[22] Trying to counter the red herring, Pusztai was saying that the Con A had been used as a 'control' experiment. The crucial experiment had been undertaken with the snowdrop lectin, GNA. The error also appeared in the *Express*, which splashed its headline with 'Genetic Crops Stunt Growth'.[23]

The effect of introducing Con A into the story was 'devastating', says Pusztai. This was because it had 'planted into peoples mind that we were doing experiments with genetically modified potatoes into which the gene for a toxin had been inserted and therefore why were we so surprised that it had caused adverse effects to the rats.'

The myth is still used to counter Pusztai today. It was used to undermine Pusztai even after it had been established that the experiments had in fact been done with the snowdrop lectin, GNA. Jack Cunningham, the Government Enforcer repeated the misinformation, saying 'it is simply not sensible to conclude that if a laboratory experiment with a known toxin added caused damage to rats that all other GM products are therefore unsafe.'[24]

Writing the same month, Professor Derek Burke, chair of the ACNFP from 1988 to 1997, sought to dismiss Pusztai's experiments: 'When he [Pusztai] added a substance called Con A to the potatoes, the animals went sick because their immune systems were damaged. That wasn't at all surprising since Con A is known to be poisonous. And if you add cyanide to lemonade you won't be surprised if it kills you when you drink it.'[25]

The 150 seconds that changed the GM debate

So what exactly did Pusztai say? On Monday night, *World in Action* broadcast. 'We're assured that this is absolutely safe,' said Pusztai. 'We can eat it all the time. We must eat it all the time. There is no conceivable harm, which can come to us. But as a scientist looking at it, actively working in the field, I find that it's very, very unfair to use our fellow citizens as guinea pigs. We have to find guinea-pigs in the laboratory.'

Dr Pusztai had been told not to talk about his experiments in detail, but he did say, in a sentence that would become the centre of the controversy, that 'the effect was slight growth retardation and an effect on the immune system. One of the genetically modified potatoes, after 110 days, made the rats less responsive to immune effects'.

He continued: 'If I had the choice, I would certainly not eat it till I see at least comparable experimental evidence which we are producing for our genetically modified potatoes. I actually believe that this technology can be made to work for us. And if the genetically modified foods will be shown to be safe, then we have really done a great service to all our fellow citizens. And I very strongly believe in this, and that's one of the main reasons why I demand to tighten up the rules, tighten up the standards.'[26]

'I still think what I said was common sense', says an unrepentant Pusztai now, 'and I wouldn't like to withdraw any of it. We are eating something that is untested. It stands to reason that there are risks and these risks we ought to find out before and not after. There is nothing inflammatory about that'.

On the evening of the broadcast, Professor James 'congratulated,' Pusztai on his TV appearance, commenting on 'how well Arpad had handled the questions'. After all the hype surrounding the day, 'what he said I thought was modest,' says James.

The following morning a further press release from the Rowett noticed that a 'range of carefully controlled studies underlie the basis of Dr Pusztai's concerns'.[27]

The riddle of the Rowett

Once again Pusztai and the Rowett differ in what happened next. On Tuesday James maintains he asked Pusztai's staff for the GNA data for the 110-day experiment, which he claims they told him did not exist. 'I couldn't believe it, because he had told the world that he had GNA studies for 110 days on immunology and growth, which he didn't have', says James, who adds that the Con A data did not exist either. 'I just said that this is the end of the world for us all'. James maintains that this is the reason why Pusztai was suspended on the Wednesday.

James' version of events is contradicted by Pusztai and other evidence. On Tuesday afternoon, the day after the programme, Pusztai's wife, Susan Bardocz, wrote a summary of all the work that had been undertaken by her husband's team. This was a five-page document that read:

'During the 3 years of the programme we conducted several experiments with the parent potato lines spiked with ConA or GNA (in amounts expressed in transgenic plants) as these were the two proteins whose genes were favoured by plant breeders in our collaboration. In addition, we carried out some short- and long-term experiments using transgenic potatoes with the GNA-genes inserted into their genomes. The responsiveness of the immune system was also checked'. The long-term experiment was number D237.[28]

This document clearly shows that long-term GNA feeding studies had been undertaken, including checking for any kind of immune response. It was given to Philip James, Ian Bremner, the Rowett's deputy, Andy Chesson, at a meeting with Pusztai, Susan and the team's immunologist at 3.00 pm on that Tuesday afternoon. Pusztai maintains that it was agreed that Ian Bremner would use Susan's report to write an authoritative press release to be released the next morning, clearing up the controversy over Con A and GNA lectin and backing up Pusztai.

But on Wednesday morning, instead of releasing the press release written by Ian Bremner, Pusztai and Susan were told to hand over their data. All GM work was stopped immediately and Pusztai's team was dispersed. His three PhD students were moved to other areas. Pusztai was also removed from other European work. He was threatened with legal action if he spoke to anyone. His phone calls and emails were diverted. No one was allowed to speak to him either. 'Even my personal assistant was told not to talk to me. I was then in limbo. I couldn't go into the lab, but they couldn't throw me out as I still had four months of my contract to work', he recalls.

Professor James also announced that he was setting up an audit of Pusztai's work, in order to 'clear' Pusztai's name. Pusztai was told to hand over everything to the Audit Committee. It was also announced that once the audit was finished, Pusztai's contract would not be renewed. It was whilst reading about his retirement in the press release on Wednesday that the enormity of what had happened finally hit home.

His 'retirement' would delight the biotech industry: 'We are very pleased that the Rowett Research Institute has publicly regretted the tremendous harm caused by publicising this type of very misleading information in the name of science', said Dan Verakis, whom Pusztai had debated with two days earlier.[29]

'Scientist's potato alert was false', ran *The Times*.[30] 'Got it Wrong', ran the *Mail*.[31] 'In one of the most embarrassing admissions by a scientific institution in years, his superiors announced that the biologist – a world authority in his field – had been talking about the wrong potatoes,' wrote Tim Radford in *The Guardian*.[32]

The Rowett press machinery was adopting Orwellian overtones and beginning to change the official story. First of all they said that Pusztai had got muddled with the wrong potatoes, then they had said that the experiments had not been done, but finally they reported that Pusztai had done the right experiments but the results were not ready yet. So the press release on that Wednesday morning said that 'by late yesterday it emerged that the relevant data provided by Dr. Pusztai referred not to experimental studies on potatoes with transgenic Con A but to GNA transgenic potatoes. The detailed analysis on the transgenic GNA studies are due to be completed by Friday'.

Although the Rowett's version is contradicted by the evidence contained in Susan's report and later by its own audit, James was sticking to his story. 'There was a series of GNA genetic modification studies that had been conducted, including the feeding studies on 110 days. My hunch is Pusztai had the body composition data for the rats for 110 days, what he didn't have was the immunology, which was the key story he was talking about.'

James appeared on the media that day to say that Pusztai 'unfortunately had muddled up two experiments' and that 'the wrong gene was identified as being involved'. The Rowett stuck to this line until the summary of the audit report showed that the studies had in fact been carried out. James also said that the reason the Rowett had allowed Pusztai to speak was that it would have been improper for the 'Rowett to be seen in any way to suppress a scientist with an international reputation who has this concern,' over his experiments.

But Pusztai had undertaken both short- and long-term feeding experiments with the snowdrop lectin, as Susan's report showed, and had immunological data. Stanley Ewen, had worked with Pusztai on the short-term feeding experiment. 'It was April '98 that the actual experiment was done, with the snowdrop lectin', says Ewen. 'All the experiments were extremely high quality, the ultimate in slickness. But we didn't expect any differences, this was only a test to develop a test to confirm the safety of these things.' Meanwhile the long-term feeding of 110 days had been started on 29 January 1998 and finished on 18 May.

Other disputed events happened on the Tuesday too. Two phone calls, Pusztai says he was told, were put through to James from the Prime Minister's office. One was 'around noon, the other was slightly earlier'. He learnt this information from two different employees at the Rowett, who could be sacked if their identities were known. The Pusztais were also later told by someone at the Rowett, currently in a senior management position at the Institute, that Bill Clinton had phoned Blair and told him to sort out the problem. 'That was the beginning of all the trouble – Arpad was sacked as a consequence of what was said in those phone calls,' says a friend.

The events of August 1998 have always puzzled Stanley Ewen, too. He often wondered what caused the sudden turn-around at the Rowett. Speaking about the incident for the first time now he is retired from the University of Aberdeen, he confirms the Pusztais' stories, but crucially he was told by yet another senior member of the Rowett. This makes four separate Rowett personnel who have spoken in private about the phone calls. 'On Tuesday, Blair phoned the Rowett twice, although everybody denies it', Ewen says.

He recounts the story of how he found out. There was an orthopaedic fundraising dinner taking place in Aberdeen. 'I didn't want to go to it at all, but it was put to me that it was a charity thing to raise money for orthopaedic research. It was put to me by one of my PhD students that I was one of her father figures, so she was putting pressure on me, and I said I would go'. Ewen attended the dinner with his wife. It was on the night of 24 September 1999. As one of the VIP guests, he was seated on the top table, next to another senior person from the Rowett. 'It was a very pleasant dinner indeed', he recalls, but it just so happened that his PhD student's 'other "father figure" was this person from the Rowett. I simply stated what a terrible time it was because of all the problems at the Rowett, and how I thought Arpad had been hard done by, by being suspended and I couldn't understand it. I couldn't understand how on the Monday it was the most wonderful breakthrough and on Tuesday it was the most dreadful piece of work and immediately rejected out of hand'.

The person from the Rowett said that 'there wasn't one, but two phone calls from Tony Blair on Tuesday', the day after the day after the *World in Action* programme. 'That's how he said it … It is my [Ewen's] understanding that the phone call was from Monsanto to Clinton, Clinton to Blair and then Blair to the Rowett.'

'I believe he was involved in senior management', says the pathologist. 'The way the Rowett runs, like many institutions, is there is a small core group of, what they call, senior management, that will make all the running and recommendations. He must have heard from a secretary who was there at the time.'

Ewen says, when he heard this, his 'jaw dropped to the floor. 'The conversation is sealed in my memory', he recalls. 'He very quickly cleared off, very sharpish. I think he realised that he had let the cat out of the bag. That was the feeling I got. Immediately great remorse overcame him and he went off very quickly.'

'I suddenly saw it all then, it was the missing link. I understood that the Director would have to toe the line, and would have to, as was presumably suggested, "kill off the research by whatever means," and it was immediate suspension for Arpad and he was threatened that if he broke his silence he would lose his pension. They have the power to do that'.

Ewen declined to name the person, but further investigations have established that it was Professor Asim Dutta-Roy. Speaking in late 2002, from his new job at the Norwegian Institute of Nutrition Research, Dutta-Roy does not recollect the conversation at the dinner party, saying 'I think I met him at a party, but it was a long time ago'.

There remains of wall of silence from the Rowett over the issue, and the staff have been threatened with dismissal if they talk about the affair, says Pusztai. The Rowett and Philip James denied any political interference to the Science and Technology Select Committee. 'At no time', wrote the Rowett, 'has the Rowett Research Institute been under any political, industrial or other pressures.' James reiterated this, saying that the decision to suspend Pusztai was made 'totally free from any influence, at any level, whether it is political, industrial.' He also maintained that, if he had received a phone call, 'I would have ignored it'.[33]

One ex-employee who was prepared to talk is Professor Robert Ørskov OBE. Professor Ørskov worked at the Rowett for 33 years, and is one of the UK's leading experts in ruminant nutrition, being currently involved in some 20 projects across the world. Ørskov is the Director of the International Feed Resource Unit, which was part of the Rowett until June 2000, when it moved to the Macaulay Institute, in Aberdeen. He is a consultant to the UN FAO and Fellow of the Royal Society of Edinburgh. He too was told about the phone calls.

Professor Ørskov says he was told that the phone calls went from Monsanto to Clinton to Blair. 'Clinton rang Blair and Blair rang James – you better keep that man [Pusztai] shut up. James didn't know what to do. Instead of telling him to keep his mouth shut, they should have told him to say it needs more work. But there is no doubt that he was pushed by Blair to do something. It was damaging the relationship between the USA and the UK, because it was going to be a huge blow for Monsanto, if it was the cauliflower mosaic virus [CaMV] promoter which was the method they used for genetic modification'.

But Professor James is adamant the phone call never happened. 'There is no way I talked to anybody in any circumstances' he says. 'It's a complete pack of lies. I have never talked to Blair since the day of the opening of Parliament in 1997.'

Although there is no proof that phone calls ever took place, Pusztai points to other evidence about Blair and GM. It is a well-known fact that Blair had been persuaded to back GM by Clinton, leading even the BBC to remark that in the GM debate 'a question mark remains over the government's independence of pressure from Washington'.[34]

In the mid-1990s the Clinton administration was backing the biotech industry 'second to none'. One White House staff member said the 1990s were going to be the decade of 'successful commercialization of agricultural biotechnology products'.[35]

Clinton's Agriculture Secretary was Dan Glickman. He says that the attitude was 'that the technology was good. It was almost immoral to say that it wasn't good because it was going to solve the problems of the human race and feed the hungry and clothe the naked. And there was a lot of money that had been invested in this, and if you're against it, you're Luddites, you're stupid ... You felt like you were almost an alien, disloyal, by trying to present an open-minded view on some of the issues being raised. So I pretty much spouted the rhetoric that everybody else around here spouted; it was written into my speeches'.[36]

There were precedents for the Clinton administration using strong-arm tactics on European governments before. Clinton and Gore were said to have pressurized both the Irish and French governments to take GM. The Clinton Administration and Monsanto were known to have a 'revolving door' with key personnel moving between the company and the administration. For example, at one stage on Monsanto's board were Mickey Kantor, former Secretary of Commerce and one of Clinton's closest advisors, William Rucklehaus, former director of the Environmental Protection Agency, and Gwendolyn King, former head of Social Security Administration.[37]

In Europe, top Clinton aides, including US Trade Representative Charlene Barshevsky, Secretary of State Madeleine Albright, Secretary of Agriculture Dan Glickman, and Secretary of Commerce William Daley, had all made representations on behalf of Monsanto.[38] Two people who played a key role in Labour's 1997 election victory, Dave Hill and Stan Greenburg, were also employed by Monsanto. Hill had left the Labour Party after 25 years to work for PR company Bell Pottinger, but he was said to be spending much of his time acting as a media advisor to Monsanto. Greenburg, who was close to Clinton and Blair, was also acting for the biotech giant.[39]

Why were Pusztai's results so threatening?

Incredible as it may seem, Pusztai's work was the first set of independent experiments undertaken to look at the health effects of GM foods. Industry-funded research had been carried out, but very little had ever been made public. Anything that questioned the safety of GM products would seriously undermine the future of the biotechnology industry, which had close political and ideological ties at that time to both the Clinton and Blair administrations.

When Pusztai spoke out in August 1998, the new Labour administration was already beginning to shape government policy for its second term. It was looking for drivers of the economy that could be trusted to deliver the growth and hence results that Labour needed. High-tech industries, such as biotechnology, were to be the central cogs of the engine that would drive the Blairite revolution, and deliver the coveted second term. What Pusztai was saying could literally derail an entire industry and with it many of the hopes and aspirations of New Labour.

But when you look at the precise nature of the Rowett work, you begin to understand how damaging Pusztai's results could be to the biotech industry, despite the fact that it was preliminary research. Part of Pusztai's remit had been to examine whether the potatoes were 'substantially equivalent'. Substantial equivalence is the internationally recognised concept by which GM food is compared to its non-GM counterpart. It was first conceived by the OECD (Organisation for Economic Cooperation and Development) in 1993, which concluded that 'If a new food or food component is found to be substantially equivalent to an existing food or food component, it can be treated in the same manner with respect to safety'.[40]

According to the principle, certain chemical characteristics are compared between a genetically modified product and any variety within the same species, past or present. If the two are grossly similar, the GM product does not need to be rigorously tested, on the assumption that it is no more dangerous than the non-GM equivalent, which should have a long history of safe use. The biotech industry has long argued that GM produce is 'substantially equivalent' to its non-GM counterpart and therefore should not undergo rigorous health or environmental studies.

One of Pusztai's first tasks was to see whether his genetically modified potatoes – those with the GNA lectin inside – were 'substantially equivalent' to their non-GM equivalents. What the Rowett scientists found was unexpected: in one of the experiments, the transgenic GNA potatoes actually contained some 20 per cent less protein and therefore were not 'substantially equivalent' to their non-GM counterparts at all.

This was an immensely important finding, as the studies had shown that as the GM potatoes were not substantially equivalent, then they should be subject to much more stringent safety tests, maybe along the lines of pharmaceuticals. Such testing would delay the whole introduction of GM food by years or even decades, and make the entire biotech revolution commercially non-viable. The whole notion of 'substantial equivalence' continues to be a major subject of controversy within the GM debate – see Chapter 7.

Moreover, this is not the first time that researchers had experienced unexpected results with GM potatoes. Researchers at the University of Oxford had genetically modified potatoes to have low levels of the NAD-malic enzyme. Unexpectedly they created high starch potatoes. 'We were as surprised as anyone,' said the head of the department undertaking the research. 'Nothing in our understanding of the metabolic pathways of plants would have suggested that our enzyme would have such a profound influence on starch production.'[41]

Secondly, and even more worrying, were the results of the actual experiments themselves. Pusztai maintains that when the rats were fed modified potatoes the development of the kidney, thymus, spleen and gut were significantly affected. So too was the body metabolism and immune function of the rats. Pusztai also found that the size of the rats' brains were affected, too, but he did not dare publicise this fact, because he believed people would think he was being too alarmist.

So what was causing the problem? No one doubts that more research needs to be undertaken until that answer is known, but there are several potential causes. If the snowdrop lectin was to blame, this could undermine the safety of other commercial products containing lectin. The year of the *World in Action* programme, some 7.7 million hectares of crops containing another lectin *Bacillus thuringiensis* (*Bt*), for example maize and cotton, had been planted.[42] Other researchers have been finding problems with the *Bt* lectin too, as outlined in Chapter 7.

However, some scientists believe it is the genetic engineering process itself which is to blame and Pusztai and Ewen were later to allude to this in their *Lancet* publication. 'When you genetically engineer crops or food it is not a simple straightforward process, you are introducing several genes which may disrupt a whole host of other gene functions, and the whole process is very difficult to predict', says Dr Sue Mayer, from GeneWatch UK. 'Many of these issues have been neglected in the race to commercialize and there has been very little attention given to those other genes, and what effect they may be having.'

Biotech proponents often argue that biotechnology is a precise science, but it is not as straightforward as some would have you believe. Belinda Martineau worked for the biotech firm Calgene in developing the first biotech product, the Flavr Savr™ tomato. 'Plant genetic engineers at Calgene and throughout the rest of the ag biotech industry have no idea where the genes will end up in the DNA of a recipient plant ... insertion could even take place in the middle of a gene, disrupting and mutating it' she says.[43] These mutations could have unpredicted effects, which could take years to become apparent.

The inherent unpredictability of GM science was reflected in a major book published in March 2002 called *Fruit and Vegetable Biotechnology*. One of the authors, Dr Javier Pozueta-Romero, from the Instituto de Agrobiotecnologia y Recursos Naturales in Spain noted that 'In plants there is a preference for random integration of the introduced DNA, which frequently leads to the accidental inactivation of important genes and to variable and unpredictable expression of the transgene itself. In some plants, over 90 per cent of T-DNA insertions may disrupt transcriptional units leading to transformants with visible mutant phenotypes.'[44]

This disruption can lead to instability, which Dr Phil Dale, a leading pro-GE scientist from the John Innes Centre, has described as a 'headache' as the precise mechanisms of gene silencing remain unclear.[45] But the instability problem is widely recognized. 'Desirable new phenotypes created by genetic engineering of plants are frequently unstable', continues Dr Javier Pozueta-Romero. 'This genetic instability is due not to mutation or loss of the transgene but rather to its inactivation.'[46]

A 1994 paper in *Bio/Technology* highlights issues of instability due to gene silencing, where the transgene can be 'switched off'. 'While there are some examples of plants which show stable expression of a transgene these may prove to be the exceptions to the rule. In an informal survey of over 30 companies involved in the commercialization of transgenic crop plants, which we carried out for the purpose of this review, almost all of the respondents indicated that they had observed some level of transgene inactivation.'[47]

The authors also suggested that gene silencing was far more common than we are normally led to believe. 'Many respondents indicated that most cases of transgene inactivation never reach the literature.'[48] So not only is it more common, but it could cause severe problems. Pozueta-Romera suggests that silencing could 'also affect the expression of homologous host genes, a phenomenon referred to as co-suppression that can have dramatic consequences for the survival of the plant if it involves a housekeeping gene or a defence-related gene'.[49]

The nightmare problem for genetic engineers could be that the random nature of the transgene has caused the plant to grow correctly on the outside but inside the cells some unforeseen reaction has taken place with unknown consequences, caused in part by gene silencing. It is the regulators' worst nightmare – a technology they say is safe, but which now has unknown and unexpected consequences.

Some scientists believe that part of the problem could be a gene that is currently central to genetic engineering, called the cauliflower mosaic promoter, which activates the transgenic gene. Basically it works as an on–off switch for the GM process. The patent for the cauliflower mosaic virus is owned by Monsanto.[50] The patent to the booster to the cauliflower mosaic gene is owned by Diatech, a company owned by Lord Sainsbury, the UK Science Minister and one of the biggest financial backers to the Labour government. Diatech was put in a blind trust when Sainsbury became a Minister.[51]

According to Pusztai, an independent statistical analysis published by the Scottish Agriculture Statistical Service, included data on both the transgene and also the construct. He maintains that the statistical analysis shows damage caused by the construct, a part of which is the Cauliflower Mosaic Virus 35S (CaMV 35S) Promoter. A statistical table seen by the author shows 'significant' damage to the pancreas, liver, small intestine and brain caused by the construct.

If CaMV 35S was to blame, the implications for the biotech industry are severe as it is currently used in the majority of the GM foods that have been granted marketing consent by the EU. Of the 11 marketing consents given, seven contain the CaMV 35S.[52]

Dr Michael Hansen is a food specialist for the Consumers' Union in the USA. The very powerful nature of the promoter could be its problem, with severe consequences, he believes. 'The CaMV 35S promoter is used precisely because it is such a powerful promoter' argues Hansen. 'which leads to hyperexpression of the transgenes, having them be expressed at perhaps two to three orders of magnitude higher than of the organism's own genes. The CaMV 35S promoter effectively puts the transgene(s) outside of virtually any regulatory control by the host genome as the natural plant promoters for each gene allow'.[53]

The CaMV 35S can also cause gene silencing.[54] This worries scientists such as Hansen. 'Since the CaMV 35S is so strong,' he says, 'not only can it affect the introduced transgenes, it can also affect genes (either turn them 'on' or turn them 'off') thousands of base pairs upstream and downstream from the insertion site on a given chromosome and can even affect the behaviour of genes on other chromosomes. Consequently, depending on the insertion site, a gene that codes for a toxin could be turned 'on', leading to production of that toxin.'[55]

Already scientists are seeing things they thought they would never see, such as faulty gene therapy resulting in leukaemia in GM mice being caused by the strong promoters. For example, an article in *New Scientist* in 2002, noted that 'gene therapy pioneers are always worried that the viruses they use to shuttle therapeutic genes into patients might accidentally dump their cargo in the wrong spot on the chromosome. In this case the inserted DNA could ... trigger the expression of cancer causing oncogenes, or disrupt sentinel genes that guard against cancer. It seems that these fears were justified'.[56]

Stanley Ewen is also very concerned about CaMV 35S. 'I think it would be reasonably suggested that animal experiments should be done for each new vegetable if they are using the CaMV 35S promoter', which Ewen believes is a 'very, very powerful' growth promoter.

Ewen worries about GM foods that might be eaten raw as the promoter might not be broken down. 'What I am very concerned about is the next generation, the fruits and vegetables that are GM and have something introduced into them that make them insecticidal. If they use the same lectin in lettuces, for example, I would be concerned about that because these are usually eaten raw'.

'This is the thrust of our argument – it is that it is the present technology that is wrong, it is not GM food in general, it is if it uses this particular promoter'. He worries that it could be a trigger for cancer, and he worries about the interaction between the CaMV 35S and polyps that are found in people's intestines. A polyp is a growth that projects, often on a stalk, from the lining of the intestine, usually in the colon. They are important because they can turn into cancer. 'My concern is that we could be speeding up the development of cancer. The polyp could go through various steps more rapidly because the growth factor could be making it grow faster.'

Ewen argues that raw vegetables or raw fruit will not actually be completely digested in the upper gastro-intestinal tract as our enzymes would be quite ineffective against intact plant cell walls. 'The plant cell walls are made of lignen and when we look for example at the appendix, there is quite often easily identified sections, bits of vegetables present in the lumen of the appendix.'

'The only place in the body in the gastro-intestinal tract where digestion can occur is due to bacteria in the colon, now bacteria can break down the lignen of the cell wall and liberate the DNA from the nucleus,' says Ewen. His concern is that the genetically modified DNA, which is acting as a growth factor could attach itself and cross the cell membrane of the polyp. 'It could then find itself acting on the nucleus of the polyp because that is what a growth factor does,' says Ewen. 'Then the growth factor could accelerate the growth of the polyp. We understand that a polyp can go through several steps before it goes malignant, unfortunately we do not know what the biological life of a polyp is', he continues. Ewen worries that if the growth promoter accelerated the time it took for the polyp to become malignant then it would make cancer-screening programmes ineffectual.

A further worry for Ewen is that the CaMV 35S virus has a very 'close homology to the hepatitis B virus'. In areas of the world where hepatitis B is endemic and acquired at a very early stage, hepatitis B seems to be able to cause hepatoma, cancer of the liver, says Ewen. 'My concern might be because of its homology, that if you are taking a similar virus orally, you might be able to speed up the development of hepatoma. It has never been put to the test in an animal model, but it would be a difficult test to do, but not impossible.'

Ewen's concerns were backed up by experiments published in 2002 that showed that a 'significant proportion' of the transgene does survive in vitro simulations of the small bowel.[57] This and other studies 'cast doubt on the assumption that plant-derived transgenes will not transfer to the intestinal microflora because the nucleic acid will be rapidly and completely degraded by the digestive enzymes'.[58]

The Audit

It was these kind of concerns – although preliminary – that forced Pusztai to speak out. The political and economic stakes were high. So were the stakes over the future of the Rowett. With Pusztai effectively silenced, like any institution under fire, the Rowett went into siege mentality and forbade anyone from talking to the press.

Meanwhile the audit team began work behind closed doors. The team was chaired by Andy Chesson, from the Rowett Institute, Pusztai's ex-boss. Other members included Chesson's Rowett colleague Dr Harry Flint, Head of Gut Microbiology and Immunology, Professor Bourne, the retired head of the Institute of Animal Health and Professor Davies from the SCRI. Effectively the only external scientist was Professor Bourne. The officials from the Rowett would argue that the audit was 'completely independent' and 'external'.[59] 'There were two Rowett people and two independent people', says Professor James. 'None of the four people had had a clue what he [Pusztai] had been doing.'

The audit, too, was to be controversial. It was not peer-reviewed and never published. According to Professor Rhodes of the University of Liverpool, although the report was 'generally factually correct', it conclusions were 'biased to an extent that in my opinion would not be deemed acceptable if subjected to peer review as for a scientific journal'.[60]

Ironically for a nutritional institute, no nutritionist was employed on the team. In total, according to Pusztai, the audit team took only 10 hours to review three years work. 'It was put together in such a hurry, that it contained many, many mistakes' he argues. 'There was no organ weight data, so many inaccuracies, and few of the data in the audit report were primary and only in-house statistical analyses were carried out', he says.[61]

The audit report was beneficial in one way to Pusztai, as it did prove that experiments had taken place. Despite the Rowett's insistence that no long-term transgenic GNA potato-feeding studies had been undertaken, in the audit report produced by the Rowett it outlines the details of Experiment D237, which started on 29.1. 98 of '110 day duration', which included 'GNA whole cooked potatoes'. It also includes a graph of the results of that experiment, and it dates them as 23/05/98 and 23/06/98.[62] This means that the feeding experiment was completed in May and calculated in June 1998, that is, before *World in Action* was filmed.

The audit notes that at the end of the 110 day feeding period in the testing of the immune responsiveness, PHA – *phaseolus vulgaris isolectin* – and Con A were used, which are 'well recognised reagents used to test your immune function,' according to Pusztai. 'It will give you an index of how well the immune function is working'. That they looked at immune effects is also concluded by the audit report, which states that, as regards 'immune response,' the results 'obtained were, in most cases, far too variable to reach statistical significance'.

This means, responds Pusztai, not only were the tests undertaken, but that he was correct after all. 'It means that in some cases it was statistically significant, therefore I had the right to say in the TV programme that the immuno-responsiveness of the rats fed by GM potatoes was depressed. Their wording gives me full vindication', he argues.

Other scientists also believe the audit actually vindicates Pusztai. For example, Professor Rhodes, who was later to sign a Memorandum backing Pusztai, noted that the audit's conclusion that 'the results obtained, were in most cases, far too variable' 'clearly implies, by the use of "in most cases" that in at least some cases statistical significance was shown, which is clearly correct'.[63]

Rhodes went on to suggest that, contrary to the audit's conclusion that the results were 'too inconsistent to draw any meaningful conclusions', that a 'fairer' conclusion would be that the experiments 'have shown statistically significant alterations in lymphocyte function', which 'deserve further study'.[64] But rather than vindication, all Pusztai received was vilification. 'The Audit Committee is of the opinion that the existing data does not support any suggestion that the consumption by rats of transgenic potatoes expressing GNA has an effect on growth, organ development or immune function' they concluded in their summary.[65]

The Audit Committee stand by their conclusions. 'The audit report made clear that he had over-interpreted his data. I wouldn't for a moment say he was dishonest about it, but I think he over-interpreted it', says Professor Bourne.

Arpad and Susan Pusztai received a copy of the audit report in August 1998, one of only ten copies ever printed. Susan was forbidden to talk to others at the Rowett about it as she, too, finally found herself suspended. The reason was simple, she was told: 'Because you are Arpad's wife'. Susan finally retired from the Rowett on 1 May 2000 after nearly two harrowing years, and remains silenced on the subject for life.

By October 1998, the political fall-out had reached the House of Lords, with Professor James and Dr Chesson due to appear before the Lords Select Committee. Pusztai alleges that he was only given three days to respond to the audit committee. 'I said, look how can I do this because you took all our data', says Pusztai, who was then given back most, but not all, of his data.

At first Dr Pusztai produced detailed criticisms of the audit report, and after becoming so enraged at the mistakes he saw, he and Susan decided to write an 'Alternative Report'. They worked day and night for the next 72 hours. Exhausted, Arpad and Susan delivered the Alternative report to Professor James' house, only to find that he was not in Aberdeen.

Giving evidence to the Lords, James admitted that transgenic potato studies had actually been undertaken with GNA. It was 'also true that they were not done with the other lectin, the Con A, which was the subject of such intense media pressure on the first day'. James said that 'the plants used and these particular studies did not have anything to do with their putative release in to the food chain', a fact disputed by Pusztai. James then said that 'Pusztai was no longer suspended because the audit is complete', although technically the Rowett had already forced him to retire, by moving his staff to other groups and forbidding him to talk to them.

James made, albeit at different times in his evidence, two statements that could be construed as incompatible. On the one hand, he repeated the audit findings and maintained that 'there were no grounds for concern on the basis of the studies that they had looked at'. On the other, he said that 'Dr Pusztai has come out of this audit review exonerated and to be seen as we all knew him as an intense investigative scientist with an international reputation'.[66] When asked about this, James says Pusztai 'was not actually guilty in my terms of deliberately misleading the world. He did mislead the world, but he was so wrapped up in the concept, he didn't distinguish between his experimental results and what he considered was a big issue of safety of GMOs'.

By the end of 1998, the Pusztai saga could have slowly subsided, with the scientist forbidden to talk to inquiring journalists. But wherever he went, scientific colleagues were curious to find out what had really happened to their colleague. Although banned from talking to the press, he was not banned from talking to other scientists outside the Rowett. Some asked to see the audit report and his own alternative response. The ensuing correspondence led to a memorandum backing Pusztai being signed by 30 international scientists from 13 countries which was published in February 1999. The scientists said:

> *We are of the opinion that although some of the results are preliminary, they are sufficient to exonerate Dr Pusztai by showing that the consumption of GNA-GM-potatoes by rats led to significant differences in organ weights and depression of lymphocyte responsiveness compared to controls. There was also strong evidence that GNA-GM potatoes and indeed two lines of GNA-GM-potatoes in the study were also different. This makes a very strong case for the necessity of performing further work to elucidate the toxico-pathological importance of these findings.*[67]

One of the scientists backing Pusztai was the former principal scientific officer at the Rowett, Dr Kenneth Lough. 'The institute is at risk of sending out signals to scientists working in [this] field of research that any sign of apparent default will be treated with the utmost severity. The awareness will of course act as a strong deterrent to those who wish to conduct research in this vitally important field.'[68]

When the scientists' memorandum was published it once again fanned the front pages. 'Ousted Scientist and the Damning Report into Food Safety' ran *The Guardian*. 'Scientist in Frankenstein Food Alert is Proved Right', said *The Mail on Sunday*.[69]

Lost in the controversy was the simple fact that Pusztai's backers were calling for the experiments to be repeated. Both from a scientific and lay perspective it would seem much more logical to repeat experiments that have disputed results, rather than just trying to silence the messenger. The Rowett rejected the scientists' claims as misleading, though.[70]

The following month, March 1999, Pusztai received further backing when it was announced that the same GM potatoes had caused other harmful effects. One of Pusztai's colleagues, Dr Birch, from the SCRI had fed the GM potato plants, containing the snowdrop lectin to aphids. The aphids were then fed to ladybirds. The scientists found that 'ladybird fecundity, egg viability and longevity significantly decreased ... female ladybird longevity was reduced by up to 51 per cent'. The SCRI concluded that: 'These results demonstrate that expression of a lectin gene for insect resistance in a transgenic potato line can cause adverse effects to a predatory ladybird via aphids in its food chain' and went on to warn that in no way was this work a 'worst case scenario' and that, 'ideally, the transgenic crops genetically engineered for pest resistance should be monitored closely after commercial release, to check for possible longer term, sublethal impacts on the agro-environment'.[71]

Working with Birch were the Gatehouses – John and Angharad – from Durham and Newcastle University, who had fallen out very badly with Pusztai over the whole GM potato episode. In happier days, they made Pusztai the godfather of their daughter, but the relationship had turned very sour. John Gatehouse remains critical of Pusztai: 'What he was doing was alarmist and he had not eliminated a number of other possibilities that could have caused the effect he was observing. The correct control experiments had not been done. He hadn't established the effect he was observing was a result of the potato being genetically modified,' Gatehouse maintains.

John Gatehouse acknowledges he is named as an inventor on a patent for the snowdrop lectin, although 'it is not being maintained'. Working with other scientists, the Gatehouses went on to write a further paper on the effects of the snowdrop lectin on ladybirds which contradicted their earlier paper and concluded, that 'GNA does not have significant direct toxic or adverse effects on developing ladybird larvae'.[72]

Having worked on the snowdrop lectin for years, Pusztai says he was approached in early 1997 by the Gatehouses and SCRI and asked to look at the Con A, as they felt that Con A would be a better insecticide than the snowdrop lectin. 'Very reluctantly we agreed. It was always said that when you put in a toxic protein gene, it is not surprising that you got a toxic potato. What people always forget to say is that the toxicity of the Con A was established by me and I published it quite some time before the GM potato controversy. So they are referring me to my own work. I told them though it won't have a cat in hell's chance of going through the regulatory process'.

But three years after the initial controversy, Dr Birch's team published yet another study based on, amongst others, one of the same lines of potatoes that Pusztai had fed to his rats. They found that GM could 'result in unintended and unexpected modification in the level of bioactive secondary plant metabolites of potato leaves'.[73] Here was further proof that GM was not predictable and that biological processes inside plants were being affected in a way that no one had intended.

However, back in 1999 when the media frenzy was reaching fever pitch, the scientific establishment knew it had to discredit Pusztai. The Rowett Institute announced its intention to seek an independent review of his work and contacted The Royal Society, who issued its own damnation of Pusztai.

CHAPTER 6

The 'Star Chamber'

'The Royal Society decided that, in view of the high profile given by some groups to Dr Pusztai's work, it would be in the public interest for the data available to be subject to independent assessment by other experts'

Dr P Collins, The Royal Society

'Their remit was to screw me and they screwed me'

Arpad Pusztai

The Royal Society

A week after the international scientists backed Pusztai (see Chapter 5), a secret committee met to counter the growing alarm over GM. Contrary to reassurances by the government that GM food was safe, the minutes show the cross departmental committee formed to deal with the crisis, called MISC6, knew the reassurances were premature. It 'requested' a paper by the Chief Medical Officer (CMO) and the Chief Scientific Advisor (CSA) on the 'human health implications of GM foods'.

Once again a government – this time Labour – was reassuring the public that food was safe, when quite clearly they did not know. What would happen, the minutes asked, if the CMO/CSA's paper 'shows up any doubts? We will be pressurised to ban them immediately. What if it says that we need evidence of long-term effects? This will look like we are not sure about their safety'.[1]

That very same day – 19 February – The Royal Society publicly waded into the Pusztai controversy. 'The Royal Society is establishing an independent expert group to examine the issues related to possible toxicity and allergenicity in genetically modified plants for food use', read the press release. 'Eminent scientists will be asked to review published and unpublished data. Their views, reached individually and independently, will be considered by a Royal Society expert panel, which will subsequently produce a report.'[2]

The review, which was instigated by the President and Vice-President of the Society, was set up 'in response to increased public concerns arising from conflicting media reports about the potential benefits and dangers of GM foods'. Four days later, the then President of the Society, Sir Aaron Klug, told the Parliamentary and Scientific Committee that 'the use of GMO's has the potential to offer real benefits in agricultural practice, food quality and health, although there are many aspects of the technology which require further research and monitoring'. Then in a veiled warning to Pusztai, Klug continued: 'I would stress that premature, partial or selective release, or misinterpretation of unsubstantiated research only serves to mislead the general public in a complex area.'[3]

Klug was backed by 19 Royal Society Fellows in a letter published in the national press. Included on the list was Sir Richard Southwood, whose Committee on BSE was widely criticized, and Professor Roy Anderson who would be widely criticized for his un-peer-reviewed models on foot and mouth, two years later.

Other signatories included Professor William Hill and Professor Brian Heap, who was to be on The Royal Society's Pusztai working group and Peter Lachmann, who was later accused by the Editor of *The Lancet* of threatening behaviour over his decision to publish Pusztai and Ewen's paper. The letter read: 'Those who start telling the media about alleged scientific results that have not first been thoroughly scrutinised and exposed to the scientific community serve only to mislead, with potentially very damaging consequences.'[4]

The first salvo had been fired from the prestigious offices of The Royal Society in Carlton House Terrace. Formed in 1660, The Royal Society is the world's oldest scientific organization, and past patrons include Samuel Pepys, Christopher Wren and Isaac Newton. Its current Patron is the Queen. It is the pillar of the scientific establishment, the scientific equivalent of the exclusive old boy's club.

Writing 'FRS' (Fellow of The Royal Society) after your name has long been the aim of many scientists, but it is joked that it is easier to win the Nobel Prize than to be elected to this august body.[5] Nearly half of its members are from just three Universities – Oxford, Cambridge and London, known as the scientific 'Golden Triangle'.[6] 'It remains a self-perpetuating elite,' says Moira Brown, a professor of Neurovirology at Glasgow University. 'Old, white and male.'[7]

In its investigation into 'Learned Societies', the Commons Science and Technology Committee found that 60 per cent of current Fellows are over 65 and only 3.7 per cent of fellows are female. The committee also criticized the Society's 'head in the sand attitude' attitude over race issues.[8]

Ever since its foundation The Royal Society has seen itself as the ultimate arbiter of scientific truth. But the last 40 years has seen The Royal Society change its role. Until 1960, The Royal Society's publications had carried a disclaimer in every issue stating 'it is an established rule of The Royal Society ... never to give their opinion, as a Body, upon any subject'.[9] Since then The Royal Society have become the scientists 'trusted' by government – the gatekeepers of reliable knowledge.[10] It is now the official voice of UK science. But its official primary objective, is to promote 'excellence in science', and it states that it has three roles, as the UK academy of science, as a learned Society and as a funding agency.

The justification for its review of GM foods, argues The Royal Society, was Pusztai's comments on the *World in Action* programme. 'The work on which these claims were based had still not been accepted for publication in a peer-review journal in February 1999, nor had it been presented to other experts at a scientific conference.'[11]

However, here The Royal Society is both right and wrong. They are correct in saying that Pusztai and Ewen's paper had not yet been published in *The Lancet*, although it had been submitted the previous December. But they are wrong, in that in November 1998 Pusztai and Ewen had been invited, along with Philip James and Andy Chesson, from the Rowett to a scientific meeting in Lund in Sweden, where 40 senior lectin specialists and European scientists attended a meeting of the COST 98 Action Programme of the European Union.[12]

Professor Bøg-Hansen, a senior professor from the University of Copenhagen, and one of the organizers, recalls how 'neither Professor James nor Dr Chesson were willing to participate or even reply to the invitation which is all the more regrettable because during the meeting Dr Ewen presented his findings'. Bøg-Hansen notes that 'conclusive evidence that raw GM potatoes has a profound physiological effect' was 'reported by Dr Ewen'. Pusztai too gave a paper, but because he was still officially gagged, this was not included in the conference proceedings, as Ewen's was.[13]

So, by February 1999, an abstract of the Pusztai and Ewen research had been published and discussed at a scientific conference, totally undermining the very reasons given by The Royal Society as a justification of their review. 'Did they have the right to peer review an internal report?' asks Dr Pusztai. 'They have never done it before and I had never submitted anything to them. They took on a role in which they were self-appointed, they were the prosecutors, the judges and they tried to be the executioners as well. I see no reason why I should have cooperated with them in my own hanging.'

The Royal Society had already entered the GM debate a month after the screening of *World in Action* the previous August, issuing a report entitled 'Genetically Modified Plants for Food Use'. This meant the Society could 'update' their report. On 5 March 1999, the Executive Secretary of The Royal Society, Stephen Cox, wrote to Pusztai and Stanley Ewen. 'The Royal Society is carrying out a review of safety issues related to GM food', the letter read, 'in view of recent concerns regarding genetically modified food, we will be re-visiting the section of our [1988] statement that dealt with toxicity and allergenicity of GM food.'

But The Royal Society just planned to examine Pusztai's work. 'If you look at their original letter', comments Pusztai, 'it says that were going to update the original Royal Society report, but what it came down to was condemnation of Pusztai and nothing else.' This is denied, though by the Society, who maintain that the review was not about 'individuals,' but was an examination of the science and the serious issues raised in the media coverage of GM, although no other scientist was mentioned.

On 15 March Stephen Cox wrote again to Pusztai and Ewen,[14] and four days later Pusztai responded. 'We have now managed to recover all our primary data from Rowett notebooks and worksheets', he wrote. This data had been subjected to an independent statistical analysis and now formed part of a final report which in turn formed part of Pusztai's submission to the House of Commons Science and Technology Committee, although this remained confidential.[15]

The two parties are at odds over why The Royal Society never received this crucial evidence. Pusztai maintains that it was the Science and Technology Committee who insisted that this report stay confidential. 'I told them there was an updated report sitting with the Science and Technology Committee. It is really only a couple of weeks, and then this updated report could be made available to them. They simply ignored it. They obviously had an agenda to make a judgement and this might have been embarrassing because it was an independent statistical analysis'. However, The Royal Society say that the Clerk of the Committee 'assured' them that Pusztai was free to give this data to 'anyone else', so long as he had asked the committee first.

In his letter of 19 March Pusztai wrote 'I am anxious to cooperate with you on this even though the omens are not the best... I think it far more important to be right than to rush to print in advance of being able to make a balanced and true statement on this weighty matter.'[16] Pusztai then explained that he was going on holiday from 24 March until 14 April.

The day before Pusztai left for his holiday, Stanley Ewen responded to The Royal Society too. 'My respectful suggestion would be to postpone your intended meeting to permit the present tense atmosphere surrounding GM food to subside analogous to the way that the Rowett

Research Institute handled the problem initially on 12.08.98,' wrote the distinguished pathologist. 'Public anxiety is heightened at present and further media attention might be averted. For my part, I feel that I don't wish to be manipulated, in this inappropriate manner, until my findings are fully published in peer-reviewed journals.'

'On the other hand,' continued Ewen, 'I would be happy to address the main working group so that the entire effect on the gastrointestinal tract can be seen and explained. I would be able to defend myself with back ups of my photomicrographs in any forum but the photographic evidence, in isolation, would be inadmissible without expert interpretation. I presume that your proposed working group would be truly and transparently independent of biotechnology companies and that I could be represented, for example, by the President of the Royal College of Pathologists, on your working group'.

Ewen offered to send The Royal Society the scientific abstract that he had prepared the year before for the Lund meeting, the evidence that had already been presented to his scientific peers. He concluded 'I do hope that your working group will represent the "defence", as well as the prosecution'.[17] Ewen was not asked to attend any meeting or send the scientific abstract.

In reply to Pusztai's letter, Stephen Cox, gave a written assurance that: 'It is important that the Society considers all available, relevant data so that an accurate and comprehensive statement can be produced.' The letter was dated 23 March, so The Royal Society knew there was a fair chance that it would not arrive in Aberdeen before Pusztai left to go on holiday the following day.[18]

In fact, The Royal Society had begun sending out information to reviewers on 19 March, and just over a week after they wrote to Pusztai, the Society starting receiving its anonymous reviews back.

The first review was written by a 'physiologist with an interest in nutrition'. The handwritten review commented that: 'I have already commented on the investigations at the Rowett Research Institute and have nothing further to add... This was pioneering work, but more research needs to be done.'[19] This implies that the reviewer had already made comments about Pusztai's work and therefore should not have been a reviewer.

The second reviewer lamented that there was 'no convincing evidence of effects on growth, organ development or immune function of transgenic potatoes expressing GNA lectin'.[20] The third 'found the data impossible to referee in the usual sense since proper descriptions of methodology are missing'.[21] The fourth did state that 'the independent statistical report is of a high and reliable quality'.[22] The fifth suggested more work was needed.[23] The final reviewer felt that, at worst, the preliminary findings were 'alarmist', but also recommended further research.[24]

On receiving these reviews when he returned from holiday, Pusztai responded that 'I am afraid that The Royal Society has obviously not accepted my offer of cooperation but proceeded regardless with the review without taking up my offer of a personal input into your review process. As it is, therefore, The Royal Society's decision was to sit in judgement of me rather than trying to establish the truth'.[25]

Then, on 13 May, the Society faxed Pusztai a 'revised' review from the sixth reviewer which arrived 45 minutes before the deadline for Pusztai to comment. The Royal Society argues that it on the reviewer's 'own initiative' that the review was rewritten. Pusztai notes that some wording was changed, all for the worse. The first draft had said there were several factors that would 'urge caution in the interpretation of the data'.[26] The revised draft noted these factors 'urge that the data are unsafe'.[27]

The day before, 12 May, Pusztai had written to The Royal Society saying it was 'totally inappropriate' to peer-review internal reports from the Rowett. He asked that 'in view of the incomplete nature of the information provided to your reviewers their comment so far obtained will be held in suspense and not be published.'[28]

But it was to no avail. On 18 May The Royal Society issued its damning verdict against Pusztai, at a press conference. The report said that Pusztai's work was 'flawed in many aspects of design, execution and analysis and that no conclusions should be drawn from it'.[29] In its first paragraph, The Royal Society said 'we have reviewed all available data related to work at the Rowett Research Institute'. At best this was misleading, because The Royal Society knew that other data was available. Their second paragraph was slightly clearer – on the 'basis of the information available *to us*', [emphasis added]. When asked about this, The Royal Society replies that: 'we reviewed all the data available to us; obviously we could not review data that was said to be forthcoming... However, the working group felt that they had received sufficient information to make an accurate assessment of the experiments carried out on GM potatoes'.[30]

The one word that mattered was 'flawed'. 'GM research "flawed in design and analysis"', said *The Guardian*. 'Experts Say Key Research is Flawed', echoed *The Daily Mail*.[31]

The fact that his work was condemned as 'flawed', contrasts, Pusztai believes, with the fact that he had had over 30 articles published in peer-reviewed journals using the same methodology and groups of international experts had backed his work. Moreover Pusztai's team had won the contract ahead of 28 other contenders, and the Scottish Office had checked his methodology. In August that year Pusztai published a paper on genetically modified peas in the *Journal of Nutrition* in the USA.[32] The referees for that paper were fully supportive, argues Pusztai. 'We have been using the same design for a long time.' But that paper concluded that transgenic peas could be used in rat diets 'without major harmful effects on their growth, metabolism and health'.[33] It was the same methodology, but different outcome.

'I think The Royal Society were rather savage and extreme because they were offended by Arpad's crude experiments as I was,' says Professor James, Pusztai's ex-boss, 'but that is different from saying that he did not have a story there'.

One food safety veteran, Professor Lacey, thought the Society's attack on Pusztai was 'absolutely grotesque. I don't know if his particular research is indicative of human hazard or not, all I know is that you cannot do experiments to find out ultimately. That is the whole key. If you cannot test a hypothesis, then you have to assume a hazard unless, over centuries of experience, you do establish safety.'

The House of Commons Select Committee

The same day, 18 May, the House of Commons Science and Technology Select Committee attacked Pusztai too. According to the committee, 'Dr Pusztai's interpretation of his research data was disputed, not only by the Rowett Institute, but also by an independent statistical analysis, commissioned by Dr Pusztai himself, which found "no consistent pattern of changes in organ weights" and which questioned the validity of the design of the experiment. Dr Pusztai told us that in his 110 day feeding trials, "no differences between parent and GM potatoes could be found". This directly contradicts his statement on *World in Action*. Dr Pusztai's appearance before us attracted far more press interest than did some of our more credible witnesses'.[34]

This is a highly damaging attack on Pusztai and one he feels is very misleading. On two crucial occasions when Pusztai tried to explain himself, the Chair of the committee cut him off, on one occasion saying that 'we do want to try to keep the science to a minimum'.[35] The Chair of the committee suggested a private meeting between Pusztai and MPs and even congratulated him afterwards, but the meeting never happened. 'You cannot argue about scientific facts without saying something scientific. They were downright unfair to me,' says Pusztai. 'But they had to destroy me so that Jack Cunningham, the Enforcer could say there was no credible evidence' against GM.[36]

The committee ignored Pusztai's submitted evidence. This concluded 'the existing data support our original suggestion that the consumption by rats of transgenic potatoes expressing GNA has significant effects on organ development, body metabolism and immune function that is fully in line with the significant compositional differences between transgenic and corresponding parent lines of potatoes'. This, contrary to the committee's report, does not contradict what Pusztai said on *World in Action*.[37]

Part of that submitted evidence is the 'preliminary' independent statistical analysis of Pusztai's work that was published by the Scottish Agriculture Statistical Service and called 'high and reliable quality' by one Royal Society reviewer. The summary states that 'differences in organ weights between rats on different diets, including those between genetically modified and parent potatoes and those between transgenically expressed GNA and added GNA, were larger and more frequent than could be attributed to chance.'[38] This sentence was omitted by the committee in its final report.

Buried deep in the statistical analysis was data that showed that there was not only a problem with the GM potatoes, but also that damage was caused by the construct or Cauliflower Mosaic Virus promoter, which caused 'significant' damage to the pancreas, liver, small intestine and brain. 'Both in the transgene and in the construct, there was statistically significant effect – this not my analysis, this is from the independent analysis' says Pusztai.

The committee did quote the sentence from the statistical analysis that said 'no consistent pattern of changes in organ weights', had been determined, as if this undermined Pusztai's work. However Pusztai argues that the committee either did not understand or chose to ignore the fact that the four different experiments were designed to look at four different things and therefore no consistent changes in organ weights should have been expected.

The statistical analysis had concerns about the designs of the experiments, a point picked up on by the committee. But then, argues Pusztai, they made a huge mistake. 'Dr Pusztai told us that in his 110 day feeding trials, "no differences between parent and GM potatoes could be found"'. This contradicts his statement on *World in Action*', they said. Actually it does no such thing, he retorts. The experiments had been designed for the rats to grow at the same rate, but in order for this to happen the rats fed GM potatoes were given extra protein. This extra protein is the measure of the growth difference.

As Pusztai explains: 'The nutritional basis of comparison is that you cannot compare anything that contains different amounts of protein or different amounts of energy or different amounts of vitamins or salt. This is the cornerstone of nutrition. The extra amount of protein that you have to add or compensate for is the measure of the difference in nutritional value. The measure of the compensation is the measure of the growth retardation'.

The Select Committee also had a confidential draft of *The Lancet* paper that had been submitted to the journal earlier in the year. Although this was later published by the UK's top medical journal with 'no significant changes', the Select Committee ignored that, too.

If the Select Committee wanted to discredit Pusztai, they achieved their aim. 'The scientist who was at the centre of the controversy of genetically modified food has been condemned in two highly influential reports published today,' said one news report. 'Research suggesting that GM foods could be a health hazard has been thoroughly discredited', ran ITN.[39]

It is beyond coincidence that The Royal Society and the Science and Select Committee published on the same day and that the ACNFP and the Committee on Toxicity of Chemicals in Food, Consumer Products and the Environment also published that week. The reason for the coordinated counter attack was simple. The Government was desperate to 'try and draw a line' under the GM affair.

Back in December 1998, the government had initiated a review of the regulatory system as it stood on GM.[40] This review had been partly scuppered by the Pusztai fall-out. Now the government had to regain the initiative. Political insiders say that pressure was put on the Science and Technology Committee and The Royal Society to discredit Pusztai, thereby enabling the government to take control again.

This behind-the-scene coordination was partly revealed by a memo showing that the government had set up a 'Biotechnology Presentation Group', which included senior Ministers. A meeting was held on 10 May when Jack Cunningham, the Cabinet Enforcer, Tessa Jowell from the Department of Health, Jeff Rooker, the Food Minister and Michael Meacher, the Environment Minister were present, and the memo is illuminating. A decision was taken to 'present the government's stance as a single package by way of an oral statement in the House. This would allow the government to get on the front foot'.

Part of the government 'package' was the paper by the CMO/CSA on GM foods that had been proposed at the secret cross-departmental meeting in February. The leaked memo revealed how 'it would not be possible to publish this paper until after The Royal Society report on the Pusztai experiment' and therefore the Office of Science and Technology 'should attempt to establish The Royal Society's intentions on timing'.[41]

This is exactly what happened. On 21 May, just three days after The Royal Society and Select Committee published – Jack Cunningham stood up in the House of Commons: 'Biotechnology is an important and exciting area of scientific advance that offers enormous opportunities for improving our quality of life,' said Cunningham. 'In agriculture, genetic modification has the potential to ensure the more efficient production of food that is more nutritious, tastes better and requires fewer pesticides. That is just the start. There are many real and exciting benefits and potential benefits, but the technology is new and the risks must be rigorously assessed.'

As had been planned, Cunningham announced the publication of a report by the CMO, Professor Liam Donaldson, and the CSA, Sir Bob May, on genetically modified food. This concluded that: 'There is no current evidence to suggest that the genetically modified technologies used to produce food are inherently harmful'.[42]

The Chief Scientific Adviser at the time, Sir Bob May, believed that 'GM is 'a vital industry of the future.'[43] May was on The Royal Society Council at the time of the Pusztai review, although The Royal Society is adamant that May 'was not involved in the review and absented himself from Council when the item was discussed'.[44] Sir Bob May later left the government to become President of The Royal Society. May is another of the influential scientists from the Zoology Department at the University of Oxford, closely linked to Sir John Krebs at the FSA and Professor Roy Anderson, whose controversial models were used in the foot and mouth crisis. Anderson and May have co-authored two books.[45]

Backed by May's and Donaldson's report, as well as The Royal Society, Cunningham laid his killer punch: 'The Royal Society this week convincingly dismissed as wholly misleading the results of some recent research into potatoes, and the misinterpretation of it', said Cunningham. 'There is no evidence to suggest that any GM foods on sale in this country are harmful. The government welcomes open, rational and well-informed debate. That is the best way to safeguard the public interest. We regret that some political, some media and other treatment of the issues has not served the public well'.

In a further move to reassure consumers, Cunningham announced the setting up of 'two new advisory bodies. The Human Genetics Commission will advise us on applications of biotechnology in health care and the impact of human genetics on our lives. The Agriculture and Environment Biotechnology Commission will cover the use of biotechnology in agriculture and its environmental effects'. He also announced the formation of a surveillance unit.[46]

If Cunningham hoped that Pusztai was buried, he was wrong. At the end of May, the medical journal, *The Lancet*, entered the fray, criticizing Pusztai's 'unwise' decision to appear on *World in Action*. But the journal said the research was necessary. In view of the 'unbridled commercial approach to genetic modification, it is perhaps not surprising that companies have paid little evident attention to the potential hazards to health of genetically modified foods', wrote the journal in an Editorial. Most spectacularly, *The Lancet* attacked The Royal Society's attack on Pusztai, as a 'a gesture of breathtaking impertinence to the Rowett Institute scientists who should be judged only on the full and final publication of their work'. In conclusion, *The Lancet* noted that the 'issue of genetically modified foods has been badly mishandled by everyone involved'.[47]

Also at the end of May 1999 the Pusztais left Aberdeen for five days. On their return they were met at the airport. 'There's been a break-in' they were told. An initial search of the house established that 30 bottles of Pusztai's malt whisky collection and some foreign currency were missing. On their second search the Pusztais realized that the bags that contained all their data from the Rowett had been stolen. This break-in was followed by a further break-in at the Rowett at the end of the year. It was only Pusztai's old lab that was broken into, the burglar managing to get past the tight security.

The Royal Society Working Group

So who was ultimately responsible for the 'impertinent' report that condemned Pusztai so hastily and readily? In May 1999, Michael Gillard, Laurie Flynn, and I began to investigate The Royal Society, working for *The Guardian*. The Royal Society did not like being under investigation itself. The Society wrote to Alan Rusbridger, *The Guardian*'s editor complaining that Michael Gillard's 'threatening tone and inaccurate analysis' was 'unacceptable' and 'inappropriate in a paper of *The Guardian*'s standing.'[48]

According to The Royal Society, 'Responsibility for our conclusions lies with the working group and with the Council of the Society… Following standard practice, the names of these referees will remain confidential. I can, however, assure you that we went to great lengths to choose people with relevant expertise and with no vested interests that could be held to affect their scientific judgement in this matter'.[49]

The Royal Society also clarified that 'We selected the team of reviewers by identifying the expertises needed (nutrition, statistical analysis, immunology etc, as listed in our published report) finding the strongest individuals with those expertises, setting aside any who had previous involvements that might be regarded as potentially distorting their judgement, and setting aside also any who had commented publicly on the Pusztai affair.' They noted that members of the Working Group were selected by the same process.[50]

The Royal Society is run by a Council of 21 Fellows, who are led by five officers, the President, who was at that time Sir Aaron Klug; the

Treasurer, Sir Eric Ash; the Biological Secretary, Patrick Bateson; the Physical Secretary, then Professor John Rowlinson; and the Foreign Secretary, then Professor Brian Heap. The council, led by these five officers, along with Pusztai's working group, were responsible for the findings. They were of course guided by the reviewers, who remained anonymous.

Despite these assurances that people involved in the decision making had not commented on the Pusztai affair, they had. Three Council officers, Professor Bateson, Professor Heap and Sir Eric Ash and one person from the working group, Professor William Hill, had already implicitly criticized Pusztai in a letter signed by 19 fellows on 22 February.[51]

Four key people involved, including the Chair of the working group, Noreen Murray, as well as Professor Heap, Rebecca Bowden and Sir Aaron Klug were all part of an earlier working group issued by The Royal Society in September 1998. Their report, entitled 'Genetically Modified Plants for Food Use' was broadly seen as being positive, with a few reservations. The chairman of the working group was Professor Lachmann, the then Biological Secretary of the Society and a known GM enthusiast,[52] who was later accused of threatening the Editor of *The Lancet* about publishing Pusztai's work.

The Foreign Secretary, Brian Heap, was on The Royal Society Working Group and the Nuffield Council on Bioethics Working Group at the same time. In May 1999, the Nuffield Council's report on GM crops concluded that: 'GM crops represent an important new technology which ought to have the potential to do much good in the world provided that proper safeguards are maintained or introduced'.[53]

Of the six members of the working group, although they may not have been involved personally with biotech companies, four were or had been employed in an Institute with biotech connections. One, Professor William Hill, was also the Deputy Chair of the Roslin Institute, famous for genetically modifying animals and for cloning Dolly the sheep who died prematurely in 2003.[54]

The Lancet

When the *World in Action* programme aired in 1998, some samples of the rat gut still lay unexamined in Stanley Ewen's laboratory at the University of Aberdeen. 'Once the programme had gone out in August, I knew that I would have to do some extra work', says Ewen, in his first in-depth interview since semi-retiring. When Pusztai had spoken about what they had seen in the lab, Ewen still had to undertake the detailed histology, because other work commitments had kept him from analysing the samples.

Ewen and Pusztai intended to go to the scientific meeting in Lund in November 1998. So, in September and October, Ewen worked on the rats and started to get results. He says he was meticulous with the experiments that were done 'blind' so not as to prejudice the outcome. He then collated and photographed the results ready to present in November. 'I realised that there was a very real difference', he says. 'I presented the results. I then refined the measurements and refined the technology yet more'.

'I thought about it for so long and made so many measurements on at least three to four occasions just to make sure I was not introducing any bias', continues the pathologist. 'Eventually when our paper was published, I was startled to find that that was the very accusation that was being made. There was no way I could bias them. I was simply reading all the measurements into a computer just using reference numbers, and once all the data was collected, then they were analysed statistically. Ultimately I repeated it five times to make sure that any bias was washed out of the system.'

'To me it was absolutely certain that there was a real difference,' Ewen maintains. 'Although it didn't necessarily happen to the same extent to every animal in the group, but you would never expect that in biology. Three or four would show a very marked growth effect and the other two wouldn't be quite so marked. Objective measurements in histology are a relatively recent thing, in the old days it was quite accepted that if you saw a difference, your judgement was never questioned. You don't measure cancer cells, you just know by your own experience that the tumour is malignant and everyone accepts it.'

'With this particular technique, it is a very exacting test. Not only did I show that the there was true elongation of the crypt, but that the number of cells increased ... even the number of cells in subsequent work which has not been published – all these parameters increased in the GM group.'

Pusztai and Ewen submitted their paper to *The Lancet*. Ewen faxed a copy of the article to the Rowett before publication, as Pusztai was still required to show them any papers based on his work there. However publication was delayed by two weeks for technical reasons. 'The rubbishing brigade had been given two weeks to do the dirty on the article. I was almost sure they would stop it,' says Pusztai.

First of all came the misinformation. 'Scientists Revolt at Publication of "Flawed" GM Study', ran *The Independent*, 'the study that sparked the furore over genetically modified food has failed the ultimate test of scientific credibility'. Written by science editor, Steve Connor, it continued that 'research purporting to show that rats suffer ill-health when fed GM-potatoes has been judged as seriously flawed and unworthy of being published by a peer-reviewed scientific journal'. Connor said that the referees were against publication. One reviewer, Professor Pickett from the Institute of Arable Crop Research (IARC) at Rothamstead in Hertfordshire, a plant scientist, was so enraged that he decided to go public, breaking the scientific rule that reviewers should remain anonymous. 'It is a very sad day when a very distinguished journal of this kind sees fit to go against senior reviewers,' said Pickett.[55]

The misinformation spread to other newspapers. *The Daily Telegraph* reported that 'rejecting advice from its reviewers, *The Lancet* is to publish the now notorious GM potato research on Friday'.[56] Even the Chief Executive of the BBSRC, the research council with responsibility for genetic engineering who is a Fellow of The Royal Society, Professor Ray Baker, said 'It is irresponsible for *The Lancet* to publish a paper which has been deemed unworthy of publication by referees'.[57]

But both *The Independent* and BBSRC were wrong. According to Dr Richard Horton, the editor of *The Lancet*: 'What Doctor Pickett didn't know and what Steve Connor from *The Independent* didn't know was that we had sent the work to six reviewers and that the clear majority were in favour of publication. Of the five technical experts four were in favour of publication, but one was firmly against, one was in favour of publication, but felt it was flawed, the others were in favour for its scientific merit'.

So of the six reviewers four were in favour of publication on scientific merit. 'A clear majority of *The Lancet*'s reviewers were in favour,' says Horton. 'Having reviewers disagree whether research should be published is absolutely normal, that's what surprised me about *The Independent* piece. That's how science progresses.'

Then Horton asks questions that should have been answered by now. 'The Pusztai data raise several new hypotheses, for example, are the changes that he has observed really harmful or are they benign? Are the changes that he has observed found in humans who have taken this particular lectin?' These questions, *The Lancet* Editor says, 'are lines of investigation that must be urgently pursued. Of itself his research does not prove one way or the other that GM foods are either harmful or safe. It does open up important avenues of investigation. All scientists should welcome that'.

Then came the 'threats'. Three days after *The Independent* article, Richard Horton received a phone call from Professor Lachmann, the former Vice-President and Biological Secretary of The Royal Society and President of the Academy of Medical Sciences. Lachmann had chaired the 1998 Royal Society Working Group report on GM Food and was a known GM supporter. He was also a consultant to Geron Biomed, which marketed the cloning technology behind Dolly the sheep, and a non-executive director of the biotech company Aprodech. In addition, Lachmann was on the scientific advisory board of the pharmaceutical giant SmithKlineBeecham.[58]

In July, Lachmann had attacked *The Lancet* and the *British Medical Journal* for 'aligning' themselves 'with the tabloid press in opposition to The Royal Society and Nuffield Council on BioEthics'. It was also, wrote Lachmann, 'disturbing and unusual for an editorial in *The Lancet* to be factually so inaccurate'.[59]

Lachmann made a bold statement saying that: 'There is no experimental evidence nor any plausible mechanism by which the process of genetic modification can make plants hazardous to human beings.' He also added that Pusztai's potatoes were 'never intended to be grown as a food crop'. What the campaign of 'vilification' against GMOs ' does to the science base and the prosperity of the UK may be serious', wrote Lachmann.[60]

According to Horton, Professor Lachmann threatened that his job would be at risk if he published Pusztai's paper, and called Horton 'immoral' for publishing something he knew to be 'untrue'. Towards the end of the conversation Horton maintains that Lachmann said that if he published this would 'have implications for his personal position' as editor. Lachmann confirms that he rang Horton but vehemently denies that he threatened him.[61]

The Guardian broke the story of Horton being threatened on 1 November 1999.[62] It was the front-page lead. It quoted Horton saying that The Royal Society had acted like a Star Chamber over the Pusztai affair. 'The Royal Society has absolutely no remit to conduct that sort of inquiry.'[63]

Edited from the final published version in *The Guardian* was a lot of detail, including a paragraph outlining The Royal Society's financial support 'from oil, gas and nuclear industries and three large biotech companies'.[64] Gone also was a quote from Rebecca Bowden, who had coordinated the peer-review process of Pusztai's work. Bowden had worked for the Government's Biotechnology Unit at the Department of the Environment, before joining The Royal Society in 1998. She talked about what *The Guardian* labelled a 'rebuttal unit'. 'We have an organization that filters the news out there', said Bowden. 'It's really an information exchange to keep an eye on what's happening and to know what the government is having problems about … its just so that I know who to put up – more of a forewarning mechanism … If we've already got a point of view then we push it.'[65]

In response The Royal Society argued that it 'was in no way involved in trying to prevent publication of the Pusztai paper' and denied the existence of a rebuttal unit. It maintains that it never tried to block *The Lancet* publication. But Horton said he had also received an email correspondence from John Gatehouse at Durham who was involved in the design of Pusztai's potatoes, which, although it was sent to him, was copied by email to Rebecca Bowden at The Royal Society. Horton was surprised at this, why was Bowden being sent copies of emails that had nothing to do with her, he thought, was she coordinating the anti-Pusztai response?[66]

On 16 October *The Lancet* printed the paper by Pusztai and Ewen. The conclusion of the Ewen/Pusztai research was that 'Diets containing genetically modified (GM) potatoes expressing the lectin *Galanthus nivalis agglutinin* (GNA) had variable effects on different parts of the rat gastrointestinal tract. Some effects, such as the proliferation of the gastric mucosa, were mainly due to the expression of the GNA transgene. However other parts of the construct or the genetic transformation (or both) could also have contributed to the overall biological effects of the GNA-GM potatoes, particularly on the small intestine and caecum.'[67]

Horton pointed out that 'publication of Ewen and Pusztai's findings is not, as some newspapers have reported a "vindication" of Pusztai's earlier claims'.[68] The journal also ran a 'safeguard' commentary that criticized Ewen and Pusztai's data for being incomplete. The results were 'difficult to interpret and do not allow the conclusion that the genetic modification of potatoes accounts for adverse effects in animals'.[69]

The Lancet also published another research letter by scientists at the Scottish Crop Research Institute and the University of Dundee, led by Brian Fenton. These scientists had also examined the snowdrop lectin, GNA. They concluded that there was evidence of snowdrop lectin binding to human white cells 'which supports the need for greater understanding of the possible health consequences of incorporating plant lectins in to the food chain'.[70] Both the Fenton and Pusztai data 'raise issues about the design of studies on safety', concluded one of *The Lancet* commentators.[71]

Horton and *The Lancet* were once again attacked for publishing the work by the biotechnology industry and The Royal Society after it appeared. Peter Lachmann called the paper 'unacceptable'. The Biotechnology Industry Association criticized *The Lancet*'s decision to publish as breaking 'unfortunate new ground for a scientific journal. Put simply, *The Lancet* has placed politics and tabloid sensationalism above its responsibility to report and assess new science'. Sir Aaron Klug, the President of The Royal Society, also joined in the criticism.[72]

In response to this criticism, especially that emanating from The Royal Society and Lachmann, Ewen and Pusztai responded by writing that: 'These methods were approved by independent statisticians. Lachmann says that the experiments need to be repeated. We would be happy to oblige. If our experiments are so poor why have they not been repeated in the past 16 months? It was not we who stopped the work on testing GM potatoes expressing GNA or other lectins or even potatoes transformed with the empty vector, which are now available. If Lachmann represents the view of the Academy of Medical Sciences on GM-food safety he should use his influence to make funds available for the continuation of this work in the UK'.[73]

But the final word was left to *The Lancet*'s editor, Dr Richard Horton, who wrote, 'Stanley Ewen and Arpad Pusztai's research letter was published on grounds of scientific merit, as well as public interest'. What Sir Aaron Klug from The Royal Society cannot 'defend is the reckless decision of The Royal Society to abandon the principles of due process in passing judgement on their work. To review and then publish criticism of these researchers' findings without publishing either their original data or their response was, at best, unfair and ill-judged'.[74]

But the threats continued: 'After the letters were published, I got feedback "by proxy",' says Ewen. 'I got a warning that I had to cut down my profile. I was never actually warned by the higher echelons of the university, but it trickled down to me, via my PhD student.' The university did not like the fact that Ewen had criticized Peter Lachmann, the President of the National Academy of Medical Sciences, and a Royal Society member. 'Because we attacked him, it was bad for any university. I was told not to be so provocative.'

The publication of *The Lancet* paper was going to have a detrimental effect on Ewen's long-term employment with the University of Aberdeen, and rather than get recognition for his work, all he seemed to get was anguish. Ewen learned that 'several groups of people had recommended I be considered' for an academic award. 'I was told that the university had no interest.'

'I felt that I had done so much work that had been unacknowledged', says the pathologist. 'I felt that I deserved some recognition, but this was being blocked at a very high level by other spokespersons. It wasn't helpful to my career. When you do these sorts of things it is very difficult for your pension. Because that is what it comes down to in the final analysis: money'. Eventually he felt that he had no option left and Ewen retired on the 26 March, 2001. He now works as a consultant to the NHS. 'I wanted to draw two lines under what I considered to be a hostile situation. Whereas instead of saying well done for getting a paper into *The Lancet*, for being at the cutting edge of this debate, I thought that I was being pilloried. It came to the point that I thought they would have regarded me more highly if I had never done this work.'

Ewen was also subjected to playground politics by the UK scientific establishment. He recalls a WHO meeting in early 2000 that he attended to discuss GM foods. 'It was an empty room and I put my briefcase down on the table, but I didn't know who I was sitting next to, and it turned out that it was a member of the ACNFP. When he realised he was sitting next to me, he moved'.

The Royal Society revisited

In February 2002, the latest Royal Society report on GM crops was published, marking a much more cautious line being taken by the Fellows. 'British Scientists Turn on GM Foods', ran *The Guardian*.[75] It was effectively a U-turn from their previous positions.

Jim Smith, who had sat on the Pusztai Working Group, chaired the latest Working Group. Whilst known GM proponents sat on the committee, there were also known GM sceptics. The scientists warned that, although in their opinion, there were no known health effects from existing GM foods on the market, 'it is possible' they argued 'that GM technology could lead to the unpredicted harmful changes in the nutritional status of foods'.[76]

The scientists argued that it was the 'vulnerable' populations who might suffer most: babies, people prone to allergies, pregnant and breast-feeding women, people with chronic diseases and the elderly. They recommended that improved safety testing within the European Union was needed before more GM crops were approved. There also needed to be a tightening of regulations, especially with respect to potential allergies and also GM baby foods.

Deep in the report was a paragraph on Pusztai. It read: 'In June 1999, The Royal Society published a report, review of data on possible toxicity of GM potatoes, in response to claims made by Dr Pusztai (Ewen and Pusztai, 1999). The report found that Dr Pusztai had produced no convincing evidence of adverse effects from GM potatoes on the growth of rats or their immune function.' The Fellows continued: 'It concluded that the only way to clarify Dr Pusztai's claims would be to refine his experimental design and carry out further studies to test clearly defined hypotheses focused on the specific effects reported by him. Such studies, on the results of feeding GM sweet peppers and GM tomatoes to rats, and GM soya to mice and rats, have now been completed and no adverse effects have been found (Gasson and Burke, 2001).'[77]

In the first sentence, The Royal Society referenced the article published by Pusztai and Ewen in *The Lancet* to the 'claims they had examined'. This is not true as The Royal Society's report was published months before *The Lancet* article and when The Royal Society had only reviewed part of Pusztai's data. Moreover, The Royal Society went on to argue that feeding studies with GM sweet peppers, GM tomatoes and GM soya 'have now been completed' and 'no adverse effects have been found'. They referenced this to a paper by Gasson and Burke. This is not a primary research paper but an 'opinion' piece written by two pro-GM scientists, Michael Gasson and Derek Burke, and published in *Nature Reviews Genetics*.[78]

Gasson and Burke outline their argument as to why it is safe to eat GM foods: 'US citizens eat GM soya without any detectable effect on their health... Many consumers eat GM foods. No significant adverse effects have yet been detected on human health'. The absence of evidence rather than the evidence of absence is being used to argue that something is safe. What do they mean by 'detectable' or 'significant' and notice the 'yet been detected', implying that there might be health effects in the future?

But the situation becomes even more hypocritical if one looks at the supporting evidence. The Royal Society report cites two studies to dismiss Pusztai's claims, and these are referenced to the Gasson and Burke paper. One group fed 'transgenic sweet peppers and tomatoes to rats', the other fed 'GM soya to mice and rats, with no adverse effects'[79]

The reference for the sweet pepper and tomatoes study is 'Chen, Z-L et al, Safety assessment for transgenic sweet pepper and tomato (submitted)'. The last word – submitted – is important. It means a paper has been submitted to a journal for publication. The crucial fact is that it has not yet been published, it is not peer-reviewed and it may never be published, if the peer reviewers advise against publication. So The Royal Society were attacking Pusztai on unpublished and un-peer-reviewed work, two years after they condemned him for speaking to the media without publishing peer-reviewed work.

The hypocrisy of The Royal Society angers Pusztai. 'They had to say something and they thought they would get away with it.' Pusztai argues that the other reference quoted by The Royal Society was invalidated because the rats were basically starving. In a peer-reviewed article published in *Nutrition and Health* Pusztai outlined his critique of the paper. 'Although the design of this long-term study was acceptable its execution was poor', wrote Pusztai. 'Thus, the growth of rats was unacceptably low and only amounted to just over 20g over 105 days and the growth of mice was zero.' 'In fact,' concluded Pusztai, 'this study gave a good example of how under starvation conditions most physiological/metabolic/immunological parameters could become unreliable.'[80]

But The Royal Society dismisses Pusztai's concerns and has shifted its opinion on what is now deemed acceptable science. 'The Royal Society recognizes how important it is that research scientists should expose new research results to others able to offer informed criticism before releasing them into the public arena. The unpublished work of Professor Chen has been discussed at international conferences (for example the 6th International Symposium on Biosafety of GMOs, July 2000, Saskatoon, Canada) and we look forward to this work being published.'[81]

When asked about this discrepancy, The Royal Society confirms that although the Chen work remained unpublished, 'it had been discussed at international scientific conferences'. Ironically by their own definition, the Pusztai and Ewen research is validated. The Society also calls Gasson and Burke's work 'primary research,' although it is a literature review – no lab work was undertaken. 'An opinion article under normal circumstances could not be taken as primary research,' argues Dr Tom Wakeford, a Royal Society funded researcher from the University of Newcastle.

The highly selective use of references also angers Dr Stanley Ewen, who wrote to The Royal Society. He pointed out to The Royal Society that a paper published in *Natural Toxins* in the autumn of 1998 showed potential evidence of harm from eating GM food.[82] The paper concluded that 'mild changes are reported in the structural configuration of the ileum of mice fed on transgenic potatoes' and 'thorough tests of these new types of genetically engineered crops must be made to avoid the risk before marketing'.[83]

'Hence I remain concerned', says Ewen, 'that ingestion of raw GM foodstuff by humans could accelerate the polyp cancer sequence in the colon as the uncooked plant cells wall will only be broken down by colonic bacteria'.

But the fundamental flaw in the scientific establishment's response is not that they try and damn Pusztai with unpublished data, nor is it that they have overlooked published studies, but that in 1999, everyone agreed that more work was needed. Three years later, that work remains to be undertaken. Pusztai maintains that you 'cannot by any stretch of the imagination, call it [the Chen paper] a repeat of our experiments.'

A scientific body, like The Royal Society, that allocates millions in research funds every year, could have funded a repeat of Pusztai's experiments. Is it that it is easier to say there is no evidence to support his claim, because no evidence exists, than it is to say that no one has looked?

By trying to sound more cautious on GM, the Society tied itself in a mess. On the one hand, the scientists argued that there was no evidence of any harm caused by GM crops, but on the other The Royal Society Fellows took issue with the very safety mechanism – substantial equivalence – designed to make sure products were safe. But it had to act. Since the late 1990s public concern about the health and environmental effects of GM crops in the UK had become more intransigent, a fact reflected by the Fellows, who recognized 'public concern' about GM technology.

But it was their international peers that had forced the shift in thinking. A year earlier, in February 2001, the Royal Society of Canada had published a report into GM, that had castigated the notion of substantial equivalence. Their Expert Panel recommended that 'the primary burden of proof be upon those who would deploy food biotechnology products to carry out the full range of tests necessary to demonstrate reliably that they do not pose unacceptable risks,'[84]

'When it comes to human and environmental safety', the Chairman, Conrad Brunk of the University of Waterloo, said 'there should be clear evidence of the absence of risks; the mere absence of evidence is not enough'.[85]

As we shall see in the next chapter, our regulatory environment is one where absence of evidence seems to have taken precedence over evidence of absence. The Royal Society believe that 'there is no reason to doubt the safety of foods from GM ingredients that are currently available, nor to believe that genetic modification makes food inherently less safe than their conventional counterparts'. But how true is this?

CHAPTER 7

Stars in their Eyes

'I have always thought that what will become the biggest problem is something we haven't thought of; there is likely to be an unpredicted impact somewhere because of the nature of GM technology and the way in which it is being used. There could be the horrible effect of someone dying because they can't avoid it'

Dr Sue Mayer, GeneWatch UK

'The only definitive test for allergies is human consumption by affected peoples, which can have ethical considerations'

Louis Pribyl, FDA scientist[1]

On Monday 10 August 1998 Dan Verakis, chief spin-doctor for Monsanto UK, debated Dr Pusztai on *GMTV*, about the *World in Action* programme that evening. Along with trying to discredit Pusztai's research, he reassured viewers that GM technology was safe. 'I think to say we are guinea pigs is wrong', said Verakis. 'These crops have been tested through 25,000 different field trials around the world. 45 different crops and about 45 different countries around the world ... No one has raised any serious issues that there will be problems with these foods.'[2]

Contrary to Verakis' assertions, biotech companies do not carry out extensive heath and safety testing – what they do is undertake as much research as is necessary to get the product through the regulatory process.[3] The majority of studies undertaken by companies remain un-peer-reviewed or unpublished.[4] Many are not even undertaken on the actual GM plant, just an isolated protein.[5] To date, there is not a single peer-reviewed human-subject study that demonstrates the safety of GM food.[6]

The testing may not tell you anything anyway. 'It is quite possible that you can have no untoward affect on test animals and you could have a significant effect on the human population,' argues Professor Richard Lacey, who cites the case of thalidomide, where the agent had no effect on rats, but a dramatic effect on people.

In many cases the regulatory authorities, such as in those in the USA, no longer demand safety testing. If the industry has managed to persuade the relevant authorities that their products are 'substantially equivalent' to their non-GM counterpart, then they are said to be safe. This has been the cornerstone of getting GM products approved.

The biotech industry and scientists would have us believe that the regulations are watertight and GM products are safe. 'All of our evidence, experience, and data indicate that crops and foods produced through biotechnology are at least as safe and perhaps safer than traditional foods,' says Val Giddings from the US Biotechnology Industry Organization.[7] There is 'no evidence that anybody has had even a sneeze from GM foods', said a CropGen representative in an interview in October 2002.

However, if you start to pick away at the evidence, these statements begin to unravel. In 2000 the EU-US Biotechnology Consultative Forum, stated 'There is a lack of substantial scientific data and evidence, often (presented) more as personal interpretations disguised as scientifically validated statements'.[8] The same year, an article in *Science* found a large number of opinion pieces on GM food, but only eight studies that presented experimental data. 'One of the more surprising results of this review was the absence of citations of studies performed by biotechnology companies', said the author.[9]

This comes as no surprise to Dr Chuck Benbrook, former executive director of the Board on Agriculture of the US National Academy of Sciences. 'Promoters of the technology and certainly the federal government in the early 1990s embraced biotechnology so enthusiastically that there was just no patience, no interest in, no serious investigation of those potential problems. It was sort of a 'don't look, don't see policy'. As a result, there really was no serious science done in the United States for most of the 1990s on the potential risks of biotechnology.'[10]

The USA

In the early 1980s, the US government decided that, although there was a need to regulate transgenic organisms, this could be achieved under existing regulations.[11] The overall responsibility for the safety of GM food safety fell under the US Food and Drug Administration (FDA). In 1992, the FDA decided that GM food was not inherently dangerous and was substantially equivalent to conventional food. As a result, the FDA makes no finding of safety before biotech foods are brought to market. The responsibility for making sure GM food is safe lies with the food manufacturers.[12]

The first biotech food approved for human consumption in the USA was the Flavr Savr™ tomato, manufactured by the company Calgene. One of the scientists involved in the creation of the Flavr Savr™ was Belinda Martineau, who wrote a book called *First Fruit*. The idea behind the tomato was simple. In today's multibillion dollar tomato industry, prices are heavily dependent on weather, season and spoilage. To reduce spoilage and increase their shelf life, many tomatoes are now picked green and ripened artificially by ethylene gas. Although these tomatoes may look red they still have a 'green' flavour and, therefore, a very high level of consumer dissatisfaction. Calgene hoped that an inserted gene would allow the fruit to remain on the vine until maturity instead of picking them at the immature, green stage.[13] By delaying picking the company hoped the tomatoes would have a superior taste that would delight the consumer so much that the 'Flavr Savr tomato could open the door for the entire agricultural biotechnology industry'.[14]

However, the results were disappointing. Instead of being firmer as they ripened, it turned out that the Flavr Savr gene only slowed the overripening or rotting process of the tomato. Whilst this would allow the fruit to stay on the shelves longer, it 'was not going to help Calgene or any other company sell vine-ripened tomatoes'.[15] As its scientists struggled with slow-ripening tomatoes, Calgene initiated discussions with the FDA in February 1989.

As Martineau explains: 'The first item on Calgene's regulatory agenda was not to convince the FDA that its first potential product, the tomato, was safe. Instead the plan called for demonstrating that the company's selectable marker gene, a gene conferring antibiotic resistance that was inserted into every one of Calgene's genetically engineered products, was both necessary to the process of producing transgenic plants and safe to use and consume'.[16]

Marker genes are an inherent part of the GM process and are used to identify which cells have been successfully modified from the ones that have failed. A common, but controversial, practice has been the use of antibiotic resistant marker genes. GM cells are given an antibiotic resistance gene, so when they are grown in a medium containing the relevant antibiotic, only the cells that have been successfully transformed will survive. However, the worry is that these genes might be transferred to bacteria in the guts of animals or humans, and that diseases could become resistant to antibiotics. Bodies such as the BMA say that the use of such genes is 'completely unacceptable'.[17]

To prove that their selectable market gene – known as a kanamycin resistance gene or *kan^r* – was safe, Calgene initiated a three-month deadline to 'design, carry out and document all the necessary safety experiments', which was a 'shockingly' short time, says Martineau.[18] When the *kan*[r] gene was finally deemed safe by the FDA, the authorities declared they saw no reason to limit the number of *kan*[r] genes in GM plants as Calgene had offered.[19]

In declaring the *kan*[r] gene safe, the FDA went against the advice of some of its senior scientists. One scientist from the Human Health Services, part of the FDA noted that 'The Division comes down fairly squared against the *kan*[r] gene marker in the genetically engineered tomatoes. I know this could have serious ramifications'. Supporting documents note that 'the major issue of concern from a clinical standpoint is the introduction of the gene *kan*[r] into significant numbers of micro-organisms in the general population of human microflora. It would be a serious health hazard to introduce a gene that codes for antibiotic resistance into the normal flora of the general population... the sponsor should seek an alternative gene marker, one that does not involve antibiotics used in human therapy'.[20]

Another senior government microbiologist wrote 'we cannot predict what the consequences' of 'disseminating billions of copies' of *kan*[r] will be. 'We know very little about the evolution and the requirements for dissemination of antibiotic resistance determinants', noted the scientist, concluding the benefits to be gained by using *kan*[r] were outweighed 'by the risk imposed using this marker and aiding its dissemination nation wide'.[21] Both serious warnings were ignored or overruled.

Whilst US Government scientists warned against using the *kan*[r] gene, Calgene scientists tried to demonstrate the Flavr Savr tomatoes 'were food just like any other tomato and should therefore be subject to the same regulation ... as other tomato varieties produced using more traditional methods'.[22] The company lobbied the Republican Administration to weaken the regulatory process and by February 1992 the then President Bush announced plans 'to streamline the regulatory process' for getting biotech products to market.[23] 'Industry', warned one of the FDA scientists, 'will do what it HAS to do to satisfy the FDA "requirements" and not do the tests that they would normally do because they are not on the FDA's list'.[24]

On 29 May 1992, the FDA published its 'Statement of Policy: Foods Derived from New Plant Varieties'. Its central premise was that genetic engineering was a 'continuum' of traditional plant breeding and that it was its intention to regulate GM foods within the existing statutory and regulatory framework.[25] Vice-President Quayle called the FDA's policy 'regulatory relief' for the biotech industry.[26] The FDA's head of biotechnology said that government had 'done exactly what big agribusiness has asked them to do'.[27]

The FDA's statement seemed to follow political rather than scientific advice. Two months earlier, Bush's office had sent a memo arguing that there was no perceived difference between traditional breeding and genetic engineering. It read: 'the policy statement needs to clearly state that method of production is irrelevant unless it directly affects the safety of food'.[28]

This fact was disputed by the government's own scientists. One FDA Compliance Officer called the document 'schizophrenic' as it was 'trying to fit a square peg in a round hole,' by forcing the conclusion that there was no difference between GM and traditional breeding techniques. The officer noted that GM and traditional breeding were different and 'lead to different risks. There is no data that addresses the relative magnitude of the risks'.[29]

FDA scientists believed that the two processes were different, noting that with GM 'natural biological barriers to breeding have been breached'.[30] Another commented that it was unlikely that scientists could 'reasonably detect or predict all possible changes in toxicant levels or the development of new toxins' as a result of genetic modification,'[31] One scientist, Dr Louis Pribyl of the FDA Microbiology Group, called it a '"What do I have to do to avoid trouble"-type document'. Pribyl said 'it reads very pro-industry' and 'it will probably be just a political document'.

One of Pribyl's major criticisms was due to the unexpected effects of genetic engineering – scientifically referred to as pleiotropic effects – which were causing the scientists real concern. 'Unintended effects cannot be written off so easily by just implying that they too occur in traditional breeding', he warned, 'there is a profound difference between the types of unexpected effects from traditional breeding and genetic engineering which is just glanced over in this document'. 'This is industry's pet idea,' continued Pribyl, 'that there are no unintended effects that will raise the FDA's level of concern. But time and time again, there is no data to backup their contention'. Pribyl commented that with GM 'when the introduction of genes into plant's genome randomly occurs … it seems apparent that many pleiotropic effects will occur'. These effects might not be seen, warned Pribyl, because of the 'limited trials that are performed'. Until further studies were undertaken, it would be 'premature' and 'potentially unsafe' for the FDA to 'summarily dismiss pleiotropy as is done here'.[32]

Other scientists supported Pribyl, warning that pleiotropic effects were 'unpredictable' and that the 'document does not present evidence that pleiotropic effects can be controlled'.[33] Despite the overwhelming concerns of its scientists, unexpected effects were not even listed as a safety concern related to genetic engineering.[34]

There is broad agreement across the scientific community that unintended effects are a potential problem associated with GM foods. They are one of the three 'potential' risks of transgenic plants identified by the prestigious US National Research Council along with allergenicity in food, plants and pollen.[35] Indeed, scientists at Calgene had problems with unexpected effects with the Flavr Savr tomato finding that 20–30 per cent of their tomatoes contained an additional antibiotic resistance gene.[36] They also found unexpected results when 'rats fed Flavr Savr tomatoes developed some kind of stomach lesions that the control rats didn't'.[37] The lesions were later attributed to a 'test material-related cause', being corroborated by the FDA as well as the EU's Scientific Committee on Food.[38]

However, government scientists were extremely concerned by the rat feeding studies. One staff pathologist acknowledged that there was 'considerable disparity' in the findings of erosions and lesions that had 'not been adequately addressed' by Calgene. The Pathology Branch were 'unable' to determine the cause of the erosions.[39] Another noted that the erosions 'raise a question of safety' and that the 'the data fall short of "a demonstration of safety"; or of "a demonstration of reasonable certainty of no harm" which is the standard we usually apply to food additives'.[40]

Despite the concerns of scientists, the US FDA granted approval for the Flavr Savr on 18 May 1994, ruling that the tomato 'is as safe as tomatoes bred by conventional means'.[41] Three days later the tomato was on the market.[42]

The FDA ruled that because they believed that the tomato was safe subsequent biotech foods did not have to undergo similarly thorough assessments prior to commercialization. A voluntary process was implemented, instead.[43] This was against the advice of FDA officials who had coordinated the FDA process. They had warned that 'such evaluations should be performed on a case-by-case basis, ie, every transformant should be evaluated before it enters the marketplace'.[44] Even Dr Martineau from Calgene argues that the Flavr Savr 'should not serve as a safety standard for this new industry... Safety assessments of these products need to be carried out on a case-by-case basis'.[45]

Despite the groundbreaking regulatory approval, the Flavr Savr was not a commercial success as it was a 'back end' product – it slowed the rotting but did not enhance the ripening process.[46] The product was eventually recalled and Monsanto swallowed Calgene in 1997. A year later, the Alliance for Bio-Integrity filed a lawsuit against the FDA, arguing for mandatory safety testing and labelling of all GM foods. During the ensuing legal action, the FDA handed over 44,000 pages of documents to the plaintiffs' attorneys.

Dr Arpad Pusztai has examined these original Flavr Savr documents. 'Tomatoes cannot be tested as a nutritional food,' he argues, because it is a fruit and does not have enough protein or energy. 'You can give it as a supplement to the diet, but then you are looking at the effect of the supplement input and intake'. In two peer-reviewed articles, Pusztai argues that 'the tomatoes used in the different experiments were from different locations and harvested at different times'. This 'should have at least put a serious question mark to the validity of the compositional comparisons on these studies and to the substantial equivalence of the GM and non-GM tomatoes.[47]

Pusztai calls into question the rat feeding results, arguing that the findings were invalid,[48] believing you cannot dismiss the erosions found in the rats. 'In human pathology glandular stomach erosion can lead to life-threatening haemorrhage, particularly in the elderly and in patients on non-steroidal anti-inflammatory agents.'[49]

Despite these concerns, the FDA has approved over 50 GM foods, the main ones being corn and soya, although some products have already been removed from the market.[50] But many people's concerns about the potential health effects of GM foods and the lax regulatory environment in the USA were borne out by the StarLink scandal.

StarLink™

Although the primary responsibility of GM foods falls under the FDA, GM plants that produce their own pesticides fall under the remit of the Environmental Protection Agency (EPA), because the EPA bears the primary responsibility for ensuring the safety of pesticides. In 1995, the EPA granted a time-limited registration for commercial planting of one of the most important 'pest-protected' plants, called *Bt* crops. *Bt* refers to *Bacillus thuringiensis*, a soil bacterium that naturally produces crystalline (Cry) proteins that are toxic to certain insects.

Bt has been used for more than 50 years by both conventional and organic farmers as a safe form of pest control. It is sprayed onto the plant, but normally loses its effectiveness within a few days. However, with GM the gene is inserted into plants to give them their own insect resistance. But there is a crucial difference between its traditional use and GM. In organic farming, *Bt* is sprayed only periodically, whereas the transgenic *Bt* produces the toxin the whole time it is growing, which could lead to insects developing resistance, a phenomenon that has, in fact, already started to happen.[51]

In April 1997, the EPA received an application from the biotech company Plant Genetic Systems, to register a corn variety called

StarLink™ that contains a particular Cry protein, known as Cry9C. The biotech company requested that StarLink could be used in both animal feed and for human use. However, there were lingering worries over the allergenicity of StarLink, and so in a compromise deal, the EPA accepted Plant Genetic System's later request for a temporary 'split' exemption, which would mean that the corn could be fed to cattle but not to humans. It was granted in May 1998, and by two years later 33 StarLink varieties were available from some 15 seed companies across the USA. As part of the agreement, Plant Genetic Systems agreed to make sure that the corn never reached the human food supply. Plant Genetic Systems was later bought by AgrEvo Company, which merged with Rhone-Poulenc to form Aventis, whose agricultural section has been acquired by Bayer.[52]

The fears over allergenicity are one of the major health concerns relating to GM crops and fears are concentrated on the Cry protein itself.[53] Indeed food allergenicity per se is seen as a growing problem, with over 160 foods associated with some kind of allergic reaction.[54] It is normally the protein in food that causes an allergy.[55] In the USA, food allergies afflict 2–2.5 per cent of adults and 6–8 per cent of children, or about 8 million Americans alone, killing 150 a year and causing some 29,000 anaphylactic shocks.[56]

Scientists are very worried by allergenicity in GM foods. The Royal Society of Canada's Panel of Experts into GM identified the potential for allergens as a 'serious risk to human health'.[57] However, because our scientific understanding of food allergy is so incomplete, it is difficult for food regulatory agencies to evaluate the potential allergenicity of GM food.[58] In 2000 a conference organized by the OECD concluded that 'there remains uncertainty about the potential long-term effects of GM food on human health... Current methods for testing and toxicity and allergenicity ... leave some uncertainties and need to be improved'.[59]

The unresolved issues were highlighted by a meeting of FAO/WHO experts who met to discuss 'Allergenicity of Foods Derived from Biotechnology' in January 2001. The delegates noted that 'it was urgently needed to establish a reliable methodology to assess the allergenicity of new foods produced by the recombinant DNA technique'. They noted that the 'decision-tree', the guide for the evaluation of allergenicity, which had been the benchmark for the biotechnology industry since 1996, was basically inadequate and that a new one was needed.[60]

US government scientists are worried, arguing 'the production of transgenic foods raises two major concerns regarding allergenicity; the transfer of allergenic proteins to new hosts and the potential for proteins from organisms that have not previously been part of the food supply to become allergens'. They note that there is a 'similarity between Cry1A (b) and vitellogenin', a known allergen from eggs yolks, and that this 'might be sufficient to warrant additional evaluation'.[61] So worried were they that in 1999 the US EPA had sought public comment on the potential allergenicity of the Cry9C protein and also asked how it should assess that 'the reasonable certainty of no harm' safety standards had been met.[62]

In 2000 the combined issues of contamination and allergenicity of GM hit the headlines. In the spring, the EU announced that canola seed contaminated with GM had been discovered, leading to hundreds of acres of the crop being dug up in Sweden, Germany and France.[63] Also in the spring, the influential National Research Council (NRC) of the National Academy of Science in the US noted that the particular Cry protein in StarLink Cry9C, 'does not degrade rapidly in gastric fluids and is relatively more heat-stable; these characteristics of Cry9C raise concerns of allergenicity'.[64] Essentially the allergenicity would be caused by the Cry9C ability to resist heat and gastric juices, which would give more time for the body to overreact.

The NRC scientists were clearly worried about allergenicity and the lack of adequate testing. The NRC noted that in the regulation of plant products with *Bt* 'the emphasis has not been on detailed assessments of safety for humans.' They recommended that 'priority' should be given to the development of improved methods for identifying potential allergens in GM pest-protected plants.[65]

In September 2000, a coalition of environmental health and consumer groups, called Genetic Engineered Food Alert, independently tested Taco Bells belonging to the food giant Kraft in a grocery store in Washington. Test results indicated 'the presence of Cry9C corn, a variety of genetically engineered corn not approved for direct human consumption'. They had found StarLink in food. Friends of the Earth, a leading member of the coalition, pointed out that Cry9C was an allergen, and demanded that any products 'marketed using the Taco Bell name be immediately removed from grocery store shelves across the country'. They also demanded that the 'FDA move swiftly to test for the presence of Cry9C corn in all products containing non-organic yellow corn, the grade of corn to which Cry9C belongs'.[66]

The announcement sent shock-waves through the industry and the political establishment: 'This discovery just shows that genetically engineered ingredients should not be on the grocery store shelves when so poorly regulated by FDA', said Congressman Dennis Kucinich. 'This is a glimpse of things to come as genetically engineered products are rushed to store shelves without real mandatory safety testing and labeling programs in place.'[67]

The StarLink saga would eventually cause hundreds of products to be recalled. Cry9C was detected in up to 22 per cent of corn grain tests, with 430 million bushels of the national corn supply tainted with small amounts of StarLink. More than 28,000 trucks, 15,000 railcars and 285 barges were found to contain StarLink residues. Some 71 seed companies reported seed stock tainted with Cry9C.[68]

Numerous different foods were contaminated such as popcorn, sweet corn, noodles, soups, and beer. Then StarLink was found in white corn products too. By November 2000, the FDA issued a recall of more than 300 food products that contained StarLink.[69] A conservative estimate of the number of people exposed to the contamination Cry9C was put in the tens of millions.[70] The whole episode – costing US$1 billion[71] – is seen as a 'seminal' event in how American consumers see biotechnology.[72]

By November 2000, the EPA's key scientific expert panel met to consider the possible allergenic effects of StarLink on human health.[73] Dr John Hagelin, the Director of the Institute of Science, Technology and Public Policy testified that the 'fiasco' had demonstrated 'the shoddiness of the government's regulation, since the system failed to keep even an unapproved bioengineered crop out of our food'.[74]

In its final report of the meeting, the Scientific Advisory Panel 'agreed that there is a medium likelihood that the Cry9C protein is a potential allergen'. According to the Panel, the Cry9C protein 'met, to some degree, all' of the characteristics associated with allergenicity, but since no records of Cry9C human sensitization exist as yet, there can be no final proof that Cry9C is or is not a food allergen.[75]

Then people started to get sick. Asked by the FDA to investigate, the Center for Disease Control and Prevention concluded that: '28 people had experienced apparent allergic reactions', although later tests on antibodies were not positive.[76] This led to some of those tested, along with environmental groups. questioning the testing methods. 'Everything else I ate in the 72 hours before I got so sick, I've eaten again with no problem,' said one of the 28, Grace Booth from California. 'Frankly, I don't trust the tests.'[77]

Meanwhile, contamination was found to be much more widespread than first thought, both nationally and internationally. In the USA, grain distributors, handlers, elevators, commodity dealers and shippers were all caught up in the trouble and StarLink contamination showed up across the grain belt of the USA.[78] StarLink was found to have been fed to animals in Canada, where it was banned[79] and by October 2000 it was found in Japan, America's largest maize customer. After the discovery, Japan's consumer organizations demanded that American corn imports be stopped.[80] The following month, Friends of the Earth announced that they had found food contaminated with Monsanto's corn, which was not approved for human consumption, in Europe.[81]

For Aventis, the fiasco was a financial disaster, costing the company over US$90 million by early 2001.[82] A class-action lawsuit was filed against the company accusing it of harming American farmers through negligence. Aventis, contended the lawsuit, would be liable for all losses resulting from StarLink contamination, including losses for farmers who did not plant StarLink, but whose corn was contaminated due to cross-pollination or during grain-handling.[83]

In the summer of 2001, the EPA concluded that 'the ongoing efforts of the growers, Aventis, the federal agencies, and facets of the food industry are greatly reducing, and will essentially eliminate, by 2004 or 2005, the amount of inadvertent Cry9C in US supplies of corn'.[84] The EPA also convened another meeting of its Scientific Advisory Panel (SAP). One of those to testify was Keith Finger, an optometrist from Florida. Finger maintains he has had three allergic reactions to StarLink. The first one was four days before StarLink was confirmed in food. 'My eyes were closing. My lips were numb and swollen. I felt like I had been knocked around by Mike Tyson for a few rounds', recalled Finger, who also experienced 'horrible, horrible, itching'.[85]

The whole regulatory process was once again found wanting when one of the SAP advisors, when presented with new safety data from Aventis said due its 'severe deficiencies' it 'would be rejected on the basis of the data' for publication in a peer-review journal. Other advisors were worried about the FDA's small sample size of allergy victims and argued that it should 'get ahold of some more of those samples'. The advisors also argued that labelling of GM food should be looked at in light of the StarLink saga.[86]

Despite these concerns, two months later the EPA started evaluating whether to re-register *Bt* crops, which caused an outcry amongst consumer and environmental organizations, who castigated it for failing its regulatory duty.[87] In their submissions, the Consumers' Union and Friends of the Earth argued that key scientific studies had been overlooked. The first study, partly sponsored by the EPA, had suggested that the *Bt* proteins similar to Cry9C protein, those called Cry1Ab and Cry1Ac, could elicit antibody responses consistent with allergic reactions in farm-workers.[88]

'If Cry9C is an allergen' says Dr Michael Hansen from the Consumers' Union, 'well if that is a known allergen – all the other Cry proteins have at least 25–35 per cent similarity, since those are the same class of protein, they are going to be immediately suspect as well. There is already evidence pointing in that direction'.

The Cry1Ab toxin is currently used in Monsanto and Novartis maize, whereas the Cry1Ac toxin is used in one type of AgrEvo's (Aventis) maize. Whereas Aventis' StarLink using Cry9C had only been approved for animal feed, the Cry1A proteins have been approved for human use in the USA, although studies on Cry1Ab has been found to have detrimental adverse effects on predaceous lacewings. These peer-reviewed studies had found a 50 per cent increase in mortality, whereas, not surprisingly, studies by Monsanto had found no adverse mortality at all.[89]

The second set was a series of four studies undertaken on mice by scientists in Cuba and Mexico. They looked at Cry1Ac. In one study on mice the team concluded that Cry1Ac was a 'potent systemic and mucosal immunogen' which 'induced an intense systemic antibody response as well as the secretion of specific mucosal antibodies'.[90] A further study by the team concluded that the Cry1Ac was a potent systemic and mucosal adjuvant, as potent as the cholera toxin.[91]

'If you over stimulate the immune system, things start to go haywire, anything from allergies to asthma, to rheumatoid arthritis', argues Dr Hansen. 'Anything that affects the immune system and can affect the guts has to be carefully looked at before allowing that substance into our body.'

The following year, the team showed that the 'Cry1AC protoxin binds to the mucosal surface of the mouse small intestine' and that 'the data obtained indicate a possible interaction in vivo of Cry proteins with the animal bowel which could induce changes in the physiological status of the intestine'. Before commercialization they believed it was necessary to perform 'toxicological tests to demonstrate the safety of Cry1A proteins for the mucosal tissue and for the immunological system of animals'.[92]

Despite the concerns of their own scientists, and those of environmental and consumer groups expressed over the two-year Bt review process, in the autumn of 2001 the EPA re-registered Bt crops. Bt cotton was extended until 2006, and corn was extended to 2008. 'The health effects assessment confirms EPA's original findings that there are no unreasonable adverse health effects from these products', said the EPA.[93]

These reassurances are premature. The EPA re-registered Bt corn knowing that crucial health data has not been collated, especially that related to allergenicity. This 'data gap' is fundamental to understanding whether these GM products are allergenic or not. However, the EPA gave the biotechnology companies until March 2003 to provide data on the similarities between the amino acid sequence in Cry proteins to known toxins and allergens.[94] This extension to the industry was given five years after FDA scientists had raised this very issue as a concern that warranted 'additional evaluation'. So the true allergenicity of Bt corn remains unknown, whilst the product is on the market.

Mike Taylor is a regulatory expert who has worked for Monsanto and was Deputy Commissioner for Policy at the FDA at the time of Flavr Savr. Taylor believes that the 'allergenicity of new proteins' is a 'very important scientific issue that needs a lot of attention if this technology is going to progress the way some people would like it to'.

But there are other health issues associated with *Bt* crops that could be potentially more serious than just allergenicity. Dr Pusztai, one of the world's leading expert on lectins has always insisted that *Bt* is a lectin. When the scientific and political establishment tried to dismiss his work on potatoes, they argued that you should expect the results because some lectins were known toxins. Dr Hansen points out there is growing structural chemistry evidence that suggests that these Cry proteins are lectins, of which some are toxic. For example, research from the MRC-Laboratory of Molecular Biology at Cambridge has noted the 'striking similarity' between the structure of the *Bt* toxins and of two known lectins.[95] Researchers at Cardiff University have identified similarities too.[96] 'These toxins appear to exhibit lectin-like structures and that combined with the studies by the Cuban and Mexican scientists suggest there might be lectin-like effects, and therefore they should be more carefully looked at,' says Dr Hansen. 'Pusztai is correct.' Further research published in December 2002, showed the high correlation between transgenic and allergenic proteins.[97]

There are other worries. There is a highly toxic and cancer-causing mycotoxin called aflatoxin, that is produced in the seeds of field crops either pre- or post-harvest in some areas of the world, such as the southern US corn producing regions. In October 2000, at a workshop on toxins held in Yosemite, a group of scientists from Corpus Christi in Texas presented results of tests into *Bt* crops and aflatoxin obtained from six companies. As *Bt* crops should reduce insects such as Lepidoptera (butterflies and moths) that can contribute to aflatoxin build up, one would expect *Bt* crops to have significantly lower levels of aflatoxin than their non-GM counterparts. However, in some experiments, the scientists found the exact opposite.[98]

The scientists sought funding from the USDA to continue this research. In their grant proposal they clearly explained what these results meant. 'Tests of nine commercial hybrid pairs from six companies at two Corpus Christi sites in 1999 showed an unexpected and significantly higher (average 85 per cent) aflatoxin content in the isogenic *Bt* hybrids compared to their non-*Bt* counterparts.'[99] That is, they found more of a carcinogenic toxin in the *Bt* crops than in the non-GM. This is another unexplained result that remains unresolved, as the funding proposal was turned down. One of the crops tested – Monsanto's 'Mon810' Maize – has been granted a European marketing consent, although it can currently only be imported in a processed form such as oils.[100]

But there are other worrying developments. The 29 April 2002 edition of the *Iowa Farm Bureau Spokesman* reported how pig-farmer 'Jerry Rosman was understandably alarmed when farrowing rates in his sow herd plummeted nearly 80 percent'. Rosman could find no answer for the problem, but heard that other pig producers in a fifteen-mile radius were suffering the same problems. 'A common denominator' reported the article, 'is that all of the operations fed their herds the same *Bt* corn hybrids. Laboratory tests revealed their corn contained high levels of Fusarium mould in the *Bt* crops. When one pig producer switched back to normal corn the problem went away. 'We're working with a problem nobody has ever heard of before', said Rosman 'It's not in the book yet'.

Fusarium can produce mycotoxins, especially one called zearalenone, which is associated with pseudo-pregnancies, although this was not found in lab analysis undertaken on the corn by scientists from Iowa State University. 'So, it is still a mystery', says Gary Munkvold, an associate professor of plant pathology at the University. 'Rosman is convinced his problem is related to the Fusarium, and he is probably right. Unfortunately, it is widely acknowledged that there are unknown mycotoxins that we do not know how to detect.'[101] Another confounding factor is that it might not be the *Bt* that is the problem, but the actual chemical Roundup. Studies at the University of Missouri have found increased levels of Fusarium after Roundup applications.[102]

The environmental impact of *Bt* crops is also a contentious issue. In a study published in *Nature*, which received world-wide attention, John Losey and a team from the Department of Entomology at Cornell University undertook laboratory experiments with the larvae of the monarch butterfly. These were reared on milkweed leaves dusted with pollen from *Bt* corn. Losey's team concluded that these larvae 'ate less, grew more slowly and suffered higher mortality than larvae reared on leaves dusted with untransformed corn pollen or on leaves without pollen'.[103]

The *Nature* article was seen as evidence that GM crops could cause harm, and the biotech industry and pro-GM scientists reacted with venom. Experiments were rapidly undertaken to prove Losey wrong, although a study published by scientists from Iowa State University found similar results when monarch larvae were placed on milkweed leaves collected at the edges of *Bt* and non-*Bt* corn fields.[104] Other studies concluded that the pollen densities were too low to cause any risk to the monarch.[105] In the end, the US National Research Council concluded, 'further field-based research is needed to determine whether dispersed *Bt* pollen could have detectable effects on the population dynamics of nontarget organisms'.[106]

A further study by researchers at Ohio State University also found that *Bt* implanted in wild sunflowers increased the number of seeds, leading to fears that the plants could become superweeds. The sponsoring biotech companies were accused of blocking further research.[107] In the meantime, pest resistance to *Bt* crops has already been witnessed under laboratory conditions, and pro-GM scientists admit 'it is possible that the widespread use of *Bt* crops could lead to the evolution of several important insect pests that are resistant to the *Bt* biopesticide. This could potentially make it necessary to resort to less environmentally acceptable chemical pesticides. This is of particular concern to organic farmers because they use the *Bt* bacterium as a permitted pesticide.'[108]

In 2001, the year that 7.8 million hectares of *Bt* crops were grown world-wide, predominately in maize and cotton,[109] the *New York Times* highlighted how: 'The total number of peer-reviewed studies evaluating the safety of *Bt* corn is zero.'[110] So with the long-term health and environmental long-term effects of *Bt* crops unknown and the issues of allergenicity and carcinogenicity of transgenes unresolved, *Bt* crops could mean we use more harmful pesticides.

In Europe there remains a moratorium on the use of GM crops, but this is likely to change, and so how tight are our regulations?

Europe

Most companies that seek approval for GM crops in Europe have first sought approval in the USA, and therefore their safety dossiers reflect the procedures of the FDA.[111] These are procedures that we now know many scientists believe to be inadequate. It is no surprise that some scientists believe European regulations are both inadequate and incomplete, based on rudimentary knowledge that exposes deep flaws in the regulatory system.[112]

In Europe there have been eleven marketing consents granted for various crops and foods. We know that scientists are worried about antibiotic resistance. We know they are worried about the construct, the cauliflower mosaic virus; and we know they are worried about the Cry protein. All the marketing consents so far approved in the EU either contain the cauliflower mosaic virus (CaMV), a Cry protein or a gene that codes for resistance to an antibiotic.[113] So all have potential health problems, but can we take solace in the EU's regulatory system to make sure everything is safe?

Our first safety blanket would be that the foods had all been subject to rigorous scrutiny by publication in reputable scientific journals, but this does not seem to be the case. In 2001, Friends of the Earth published a report into testing of GM crops in Europe. They concluded that 'the majority of these tests have been undertaken by biotech companies, and very few have been published or peer-reviewed'. In all, the Friends of the Earth researchers examined 12 GM crops or GM processed oils approved by the EU for human food use. In five cases the safety tests were unpublished. In four cases the details of the studies were 'not available'.

'Biotech companies,' concluded Friends of the Earth, 'cannot expect the public to have any trust in their products if they are not prepared to expose their safety testing to independent scrutiny'.[114] So the irony is that whilst Pusztai was attacked for speaking out before peer review, our very own regulatory system – our safety benchmark – is based on un-peer-reviewed science.

In response, governments and regulators argue the regulatory process is thorough and rigorous. 'Before GM foods are approved for sale in the EU they must be rigorously assessed for safety in accordance with the requirements of the EC Novel Foods Regulation (258/97),' says the UK Food Standards Agency.

'This Regulation, says the FSA, 'which came into force on 15 May 1997, established an EU wide system, which requires that a novel food is assessed for safety before it can be released onto the market. A novel food is a food, which has not been consumed by humans to any wide extent in the EU before. This includes any food containing, or produced from, a GMO'.[115] Although this regulation is now legally binding, it means that it replaced a voluntary scheme for the assessment of novel foods which had been in operation for more than 10 years.[116]

It was a voluntary process, therefore, that the Flavr Savr went through. In February 1996, the Flavr Savr tomato became the 'first clearance of an unprocessed GM food anywhere in Europe.' Like their regulatory counterparts in the USA, the officials at the UK body that advises the government, the ACNFP, passed the tomato as safe.

The ACNFP dismissed many of the concerns we now know that FDA scientists had, especially on antibiotic resistance. 'The committee gave particular consideration to the presence of an antibiotic resistance marker gene and its gene product which inactivates the antibiotics kanamycin and neomycin. The committee was satisfied that its presence and that of its product in the tomatoes would not compromise the clinical and veterinary use of these antibiotics.'[117]

Two and half years previously, the Public Health Laboratory Service had warned that, given concerns over antibiotic resistance build-up, it would be 'illogical to sanction the deliberate release of antibiotic resistance genes via human food'.[118]

Despite this, in February 1995, UK Ministers approved the sale of three processed GM foods. They were a tomato paste made from GM tomatoes, which was manufactured by Zeneca to delay the fruit softening, an oil from genetically modified oilseed rape and processed food products from genetically modified soya beans. The press release said that the ACNFP had 'looked closely at all three processed products and agreed that they are safe'.[119]

'In the case of the tomato paste', the ACNFP argued, 'the inserted genes would have been destroyed by processing and thus would not be present in the final food.' The ACNFP concluded 'there was no consumer safety concerns'.

To back up its position, under the section 'human exposure', the ACNFP noted that 'a GM tomato developed by the American company, Calgene, and similar to Zeneca's (from which paste only will be produced) … has recently been cleared by the Food and Drug Administration (FDA) for sale in the USA. No adverse effects have been reported from people consuming the tomato'.[120]

This statement is interesting in two ways. It shows that the regulatory authorities in the UK were relying, in part, on evidence from the regulatory authorities in the USA. We know now that this was a false assurance, as the very real concerns of the FDA scientists had been overruled for political reasons. Secondly, the ACNFP said that there was no evidence of harm occurring to anyone eating the tomato. This statement is in itself a scientific solecism. What they did not mention is that no one was *looking* for harm and that there was no post-sale monitoring of people's health. This statement – that there is no evidence of harm from people eating GM food – is used repeatedly by GM proponents, but even the current chair of the ACNFP, Janet Bainbridge, admits that 'no monitoring system exists, and it seems unlikely that effective post-market surveillance could be carried out'.[121]

In their study of transgenic plants, the US National Research Council noted that it was 'non scientific' to note that there had been no harm, because there was simply no one looking. 'The absence of evidence of an effect is not evidence of absence of an effect,' it said.[122]

'There has never been any monitoring of GM crops or foods in any sense,' says Dr Sue Mayer from GeneWatch UK. 'They talked in ACNFP of setting up monitoring at one stage. I think they have decided that they didn't quite know what to do and it would all take time and money.' Indeed, the Medical Research Council (MRC) has argued that the feasibility of 'undertaking epidemiological studies to assess any post-marketing effects of GM foods on human health appears to be limited'.[123]

Mayer maintains that 'there has been no systematic effort to monitor and to check the assumptions that were made in the assessments of safety, which seems inappropriate in terms of the start of the technology'. The impacts from GM, says Mayer, 'are going to be longer term or more subtle effects which are going to take years before they become visible, such as allergies'. Although it is unlikely people will just drop down dead, 'if you look at the introduction of kiwi or peanuts, you get time lags of 5–7 years before any one begins to suspect clinically there might be a problem', she argues. 'No one is saying that GM is the only one that is going to cause health problems. But there are unique risks that may arise from genetic modification.'

Dr Mayer runs GeneWatch UK, whose remit is to 'promote environmental, ethical, social, human health and animal welfare considerations in decision-making about genetic engineering and other genetic technologies'. Mayer is at pains to point out that 'GeneWatch is not opposed to genetic technologies in principle'. But GeneWatch 'believes that public participation is crucial for robust and effective decision-making and that this can only take place in the context of openness, where debate is well informed and proper weight is attached to public concerns and aspirations for the future'. In just five years, GeneWatch has become an influential part of the genetic engineering debate, both in the UK and the EU. Mayer is a key member of the AEBC, the Agriculture and Environment Biotechnology Commission.

For nearly a decade, Mayer has been highlighting flaws in the safety testing of GM crops. Existing assessment methods often require tests for toxicity, but these are undertaken with the isolated protein and not the GM plant.[124] 'They test things in isolation', she says. 'It is this reductionist thinking that is really poor in terms of safety. A lot of the toxicity tests on soya were to do with the isolated protein and to determine whether it was likely to be allergenic or toxic, but it is taken out of context.'

'It is not a good way of looking at things' continues Mayer, 'and they still do it with *Bt* maize. They will test the *Bt* toxin separately in toxicity tests, and they will see how it gets digested in a test-tube. Of course, that is not how an animal will be exposed to it – it will be in a food where the toxin may be protected by some of the foods that won't get broken down by the acids in the stomach. All those kinds of issues get lost'.

Before GeneWatch, Mayer had worked as the Head of the Science Unit at Greenpeace, working part-time on biotechnology, where the scientists were worried about GM years before it became a public issue. 'The whole context in which judgements were made around safety in the early to mid nineties, was that there should be some safety testing done, but the underlying assumption was that it was safe and it should go ahead', says Mayer. 'The most appalling data and experimental evidence was accepted as evidence that there was unlikely to be any risk because there was this presumption that it was such a great thing to do.'

Just three months after the GM tomato paste has been approved, Dr Mayer wrote a report for Greenpeace on GM tomatoes. It was never published, but concluded that the 'risk assessments may prove to be deficient if their evaluations of the scientific uncertainties, and more importantly the framework of what is considered relevant, are mistaken'. Mayer and her colleagues concluded that there were 'several important areas in which the risk evaluation seems flawed in ways which demand a revaluation of the risks concerned'.

The report highlighted potential health impacts, that had 'centred on the use of antibiotic marker genes. Antibiotic resistance could be transferred to bacteria in the gut leading to problems in the treatment of the disease. There is also the possibility that the gene product could be toxic or allergenic'.[125]

So we know some scientists believed that the approval process was flawed, but what about the other products that the ACNFP approved? Dr Mayer is also critical of the approval process of Monsanto's soya beans that were tolerant to a herbicide that kills weeds, called Glyphosate, which is more commonly known as Roundup®. The soya was the first commodity GM crop that was imported into the EU in 1996.

Dr Mayer points out that one fundamental flaw in the experiments that the ACNFP evaluated can be seen in Monsanto's application document to the committee in 1994. This says that 'to focus the analysis on any effects of the introduced protein, the soybeans from which the seed were derived were *not* treated with Roundup herbicide' [emphasis added].[126]

'All their original safety studies were done on Roundup Ready soya beans which weren't grown with Roundup' says Dr Mayer. 'You have to remember that glyphosate acts systemically and is taken up into the plant and is in the plant. The genetic modification you have done is to change the way the plant deals with glyphosate. But that chemical may interact with the changed genome. You would just expect that any reasonable person, when you are doing safety testing, that as part of the test you would test the food as it would be expected to be grown and used.'

But the beans that were analysed to see if they were 'substantially equivalent' to their non-GM counterparts were not sprayed with Roundup either. 'The irony with that is that the soya beans that were evaluated and tested for substantial equivalence were ones that were grown without the use of glyphosate,' says Sue Mayer. 'So what was actually tested was not what was going to be grown and consumed by people. And that went through ACNFP.' Mayer believes this shows 'how sloppy they were in their thinking' and highlights the ACNFP's 'rather cavalier attitude to the risks'.

It was so 'sloppy' that Greenpeace accused the granting of Monsanto soya bean as 'regulatory negligence' by the ACNFP. Greenpeace's argument was, amongst others, that Monsanto had submitted no toxicological assessments; Monsanto had submitted no independent verification of its conclusions and no peer review of its methodology and that the data was obtained from soya that had not been treated with glyphosate. The ACNFP rejected Greenpeace's accusation, concluding that beans and derived products from GM 'were as safe for human consumption as beans derived from other conventional soya bean lines'.[127]

Mayer believes the public backlash against GM has probably made the ACNFP more sensitive, forcing them to be more thorough. But, generally speaking, argues Mayer, 'they are anxious to see the technology go ahead, so that is always going to colour their decision making'.

Even government departments are worried about 'sloppy' techniques. In 1994 Ciba-Geigy[128] applied to market GM maize that was resistant to the corn borer and tolerant to herbicides. The application was passed from the French to the European authorities for individual member states to consider. It contained the CaMV, Cry1A9b protein and *bla* gene, which is resistant to the antibiotic, ampicillin. In May 1995, ACNFP's sister organization responsible for environmental issues, the Advisory Committee on Releases to the Environment (ACRE), counselled that the product did 'not pose a risk in terms of human health and environmental safety for the United Kingdom'. ACRE had no objection to the product being placed on the UK market.[129]

However, leaked internal government correspondence shows that ACNFP could not 'recommend food safety clearance of the *unprocessed* seed' due to the *bla* gene as it could lead to new strains of ampicillin resistance bacteria. 'Apart from the potential safety hazard from the presence of the *bla* gene, Members were concerned that clearance of the unprocessed GM maize would also condone 'sloppy' genetic modification', said the letter. 'Members felt that it was not necessary for the *bla* gene to have been retained in the GM maize.' If the maize was widely marketed, the letter concluded, 'there will be a greater opportunity for the antibiotic resistant gene to spread to bacteria which may ultimately be harmful to humans'.[130]

However, MAFF wanted the advice to be returned to the European Commission to read 'the ACNFP and ACRE had 'serious concerns' about the approval, because the maize 'contains an intact *bla* gene ... We assess the risk of *bla* gene being picked up an expressed by the intestinal bacteria of an animals as relatively high'.[131]

Further internal correspondence asked 'could the Use of Ampicillin in Veterinary Medicine be Compromised? ... the answer must be yes. There are additional factors which must be taken into account and have to date been ignored. This is *not* a one-off experiment ... in effect, the ruminant is being "loaded" on a continual basis with *amp*[r] containing material.'[132] Another warned that 'the use of this product constitutes a significant risk in an era of increasing microbial antibiotic resistance and declining availability of new antibiotics'.[133]

Dr Pusztai also examined data from Ciba-Geigy, as well as data from other companies including Monsanto, having been asked to by Philip James. He noted that the 'effects of *bla* or *pat* transgenes in maize are *not* dealt with… The transgenic maize seed/plant itself has never been tested for nutritional performance, toxicology or allergenicity'. Pusztai told the authorities to 'have proper experiments with the transgenic plant and do not leave it to chance'.[134]

Eighteen months later, the European Commission's Scientific Committee for Animal Nutrition, known as SCAN, concluded that there was 'no evidence of a risk of antibiotic resistance in the bacteria of the animals digestive tract from the use of the genetically modified maize'.[135] On the same day, the European Commission's Scientific Committee for Food concluded that the transgenic maize was 'substantially equivalent to maize presently on the market'.

A month earlier, a MAFF scientist contacted by the Scientific Committee for Food had warned that the genetically modified product contained genes 'not found in nature' and there were worries of 'rapid degradation of ampicillin'. The scientist warned that 'the probability of transfer of the intact *bla* gene is low, the risk is finite. Should this occur, given the importance of the antibiotic and the high copy number of the vector, the results would be very serious'.[136]

In February 1997, the maize was approved, although that decision has subsequently been challenged legally. In September 1998, the French Counseil d'Etat ruled that the risk assessment on the insect-resistant maize was incomplete because it had failed to assess the antibiotic-resistance marker gene.[137]

The month before, the European Union granted a further marketing consent for another GM maize, this time owned by Aventis. But once again the regulatory regime was found wanting. The herbicide-tolerant forage maize, known as T25, which is marketed as Chardon LL, had been granted approval for commercial use by ACRE, back in 1996. However, any crop, whether GM or non-GM must be put on the UK National Seed List, before it can be sold to UK farmers for commercial use. A public inquiry into the Seed Listing began in October 2000. A month after opening, two scientists, Dr Steve Kestin and Dr Toby Knowles, from the Department of Clinical Veterinary Science at the University of Bristol, who have authored 146 scientific papers on related issues, gave evidence on behalf of Friends of the Earth. They reviewed the original Aventis chicken safety study that had been approved by ACRE. The Bristol duo found that in the feeding experiment, where one set of chickens ate normal maize and the others GM maize 'no statistically significant effects were found'. But there 'was a trend for higher mortality in the GM fed birds (twice as many broilers fed the GM T25 maize died (10) compared with those fed the non-GM maize (5)'.[138]

'It wasn't really a good enough experiment to base a student project on, let alone a marketing consent for a GM product,' said Kestin. Knowles was concerned that such a badly designed study that was a 'waste of time' had got past 'so many levels of scrutiny from Aventis to EU scrutiny to MAFF'. When asked how ACRE had approved the study, its then Chairperson, Professor Gray, responded that 'there is nothing in that [chicken] study, which alerts us to a safety issue'. Gray finally conceded 'the experiment should be reanalysed again'.[139]

In November 2000, the inquiry was suspended after it was discovered that official 'distinctness, uniformity and stability tests on the seeds on which the proposal to add Chardon LL to the National List was based had only been carried out for one year, rather than the two required.[140] In February 2002, ACRE held a further public hearing into Maize T25 where Aventis reconfirmed that its maize was 'as safe as its non-modified counterparts'.[141] But all the indications still are that it will be the first GM crop put on to the National Seed List. It may also be the last before the benchmark that underpins GM finally collapses.

Substantial equivalence

In its evidence to the Chardon LL Inquiry, Aventis had argued that its maize was substantially equivalent to the non-GM counterpart. In fact substantial equivalence underpinned the whole company's application,[142] as it does for all GM foods. But substantial equivalence is now a discredited and outdated safety benchmark.

Dr Mayer was one of the co-authors of a *Nature* article that criticized the concept, along with Dr Erik Millstone and Dr Eric Brunner. The three argued the concept was 'misguided' and 'vague', 'and should be abandoned in favour of one that includes biological, toxicological and immunological tests rather than merely chemical ones. Substantial equivalence is nothing more than 'a pseudo-scientific concept because it is a commercial and political judgement masquerading as if it were scientific'.[143]

In 2001, the Royal Society of Canada's panel of experts on GM attacked 'the use of substantial equivalence as a decision threshold tool to exempt GM agricultural products from rigorous scientific assessment to be scientifically unjustifiable and inconsistent with precautionary regulation of the technology'.[144] The following year, the British Royal Society also argued that 'the criteria for safety assessments should be made explicit and objective and that differences in the application of the principle of substantial equivalence, for example in different Member States of the European Union, need to be resolved'.[145]

Even the current chair of the ACNFP criticizes the very concept the committee has used to evaluate GM foods. It is 'not a sound basis for GM food risk assessment', says Professor Janet Bainbridge, 'the presumption of safety of novel GM plants on the basis of substantial equivalence lacks scientific credibility'. One of the reasons, says Bainbridge, is that the introduction of a transgene may 'cause unexpected … alterations in the phenotype of the novel organism, which the substantial equivalence approach might fail to disclose'. Although Bainbridge believes that current regulations on GM 'appear' to have protected the public 'we do not know what we may have missed'.[146]

'They are now having to rethink the whole of their safety assessment and the underlying rationale for it', says Mayer, who argues that the concept has 'crumbled as a concept intellectually'. It has particularly crumbled for what are the second-generation nutritionally enhanced foods, those 'that are being seen as the saviours of GM foods, but they are going to be a nightmare to evaluate for safety. I think substantial equivalence has had its day, you won't see the word eradicated, because there is too much invested in it, but I think what you will see are additional methods of assessment.'

So how do you evaluate safety for the next generation of GM crops, many of which are going to be eaten raw or are going to have vaccines and drugs in them? Some argue that it is impossible: 'If you had one new product, you could actually establish safety over the years to the consumer, but when you have potentially thousands of new products all at once, there is no possibility that you can establish safety,' argues Richard Lacey.

Others, like Dr Levidow from the Centre for Technology Strategy at the Open University believe our risk assessments are based on ignorance. He asks why has there been so little scientific research that could provide meaningful empirical results about toxicological or immunological effects? Why, he says, have scientific institutions emphasized the methodological limitations of studies which suggest risk – but not of those which show no harm? Whenever companies propose to market GM vegetables or fruit to be eaten in a raw form, how will any hazards be anticipated?[147]

These questions remain unanswered, as Europe faces commercialization. In a tightening of the regulations that had been proposed the year before, in June 2002, the Environment Committee of the European Parliament voted narrowly in favour of more extensive labelling of GM food and animal feeds. Although the vote was preliminary, it could mean mandatory labelling for both fresh and refined foods and the lowering of the mandatory labelling threshold from 1 per cent to 0.5 per cent. The move could also signal the banning of any products containing traces of biotech ingredients not authorized in the EU.[148]

In October 2002, the European Commission announced that the process for commercial approvals of genetically modified food and crops in Europe was to be left to biotech companies and Member States, after Ministers failed to agree legislation on labelling and traceability. Any Member State wanting to start a new commercial approval process would have to do so under the new Deliberate Release Directive, a process likely to take up to 14 months. As the de facto moratorium was effectively maintained, environmental groups called on Members States to introduce strict labelling and a traceability regime during this time.[149]

As Europe grapples with labelling and traceability laws, the StarLink fiasco serves as a warning. But the same result – that of a food not intended for human consumption – could also happen through gene flow. As *Nature Biotechnology* points out, 'gene flow (like mixing) could result in GM material unintended for human consumption ending up in the human food chain'. There are 44 plant species that could mate with wild relatives.[150] Many current GM crops 'have the potential to cross-pollinate' and this has already occurred.[151]

So it could only be a matter of time before a potentially allergenic GM trait gets passed from a transgenic plant to a non-GM plant. In the summer of 2002, the regulatory environment in the UK was again found wanting when it was announced that over 20 sites of oil seed rape planted as part of the UK field-trials contained genes giving resistance to the antibiotics neomycin and kanamycin.[152] The latter, inserted into the Flavr Savr, had given the FDA scientists great cause for concern and experts fear that both could reduce the effectiveness of medicines.[153]

Also that summer, the use of antibiotic resistance genes in particular and the safety in GM in general were thrown into question by new research from Newcastle University. Researchers found that GM DNA could find its way into human gut bacteria, and the GM DNA was taken up by gut bacteria. The experiment with seven people who had been given colostomies found that GM 'nucleic acid was detected in all seven subjects'. In three of the seven samples they found gut bacteria had taken up the herbicide resistant gene at low levels.[154] 'They have demonstrated clearly that you can get GM plant DNA in the gut bacteria', said Michael Antonio, a senior lecturer in molecular genetics at King's College Medical School. 'Everyone used to deny that this was possible.'[155]

This was just another alarm bell ringing, as is the fact that GM contamination is already happening. Once again those that raise the alarm are in the firing line. The truth is that many scientists do not believe GM crops to be safe. In 1988, US policy makers conceded that there was no way to be entirely certain of the safety of GM food. If the 'public wants progress, they will have to be guinea pigs,' said one.[156]

CHAPTER 8

Immoral Maize

'I don't want to be a martyr by any means, but I cannot avoid now realising that this is a very, very well concerted and coordinated and paid for campaign to discredit the very simple statement that we made'

Ignacio Chapela

'Current gene-containment strategies cannot work reliably in the field'

Nature Biotechnology, Editorial[1]

In the autumn of 2000 a graduate student from the University of California held a workshop for local peasant farmers in the beautiful mountainous region of Sierra Norte de Oaxaca in southern Mexico. The graduate, David Quist, hoped to show the farmers how to test their seeds for GM. To do this he thought he would show them the difference in the purity of the local maize, called criollo, compared to the maize that had been shipped in from the USA, where some 40 per cent is GM.

The US maize would test positive for GM and, naturally, the Mexican maize would be negative, he thought. But Quist was wrong. For some reason, instead of the local maize being negative, it kept coming up positive.[2]

Quist was visiting the region because his supervisor, Dr Ignacio Chapela, who was originally from Mexico City, had been working with the campesinos or peasant farmers in Oaxaca for over 15 years, assisting them in community sustainable agriculture.

Quist was told by Chapela to bring the samples back to the USA, where the two would repeat the experiments and test the native maize 'landraces' for contamination by GMOs. Although there had been a moratorium on the commercial growing of GM in Mexico since 1998, there was general concern that GM maize was coming across the border from the USA, either as seed or as 'food aid' and that it was contaminating the indigenous species.

This was seen as a worry for various reasons, the main one being that contamination threatens Mexico's unique maize genetic diversity. Mexico is the traditional home of corn, where the plant was first domesticated some 10,000 years ago. It is an important crop for a quarter of the nation's 10 million small farmers and corn tortillas are a central part of nation's diet. But now due to NAFTA (the North American Free Trade Agreement), the country is a net importer of the crop. With some 5 million tonnes coming in from the USA every year, and because there is no mandatory labelling, there is no way of knowing if this corn is GM or not.[3]

Greenpeace had launched a campaign in Mexico in January 1999 warning the Mexican Government that GM maize imports from the USA 'would end up polluting Mexican corn varieties'. 'The aim was to stop the imports', says Hector Magallon Larson, from Greenpeace Mexico. 'Greenpeace wanted to highlight the inconsistency of the Mexican government stance of supporting a moratorium but allowing millions of tonnes of GE corn to pour over the border.'

The campaign was not well received in official circles. 'The main response came from the Minister for Agriculture,' says Hector Magallon Larson. 'He said these corn imports were only for human food and animal feed, so the corn shouldn't be planted. They also said that the corn was treated with a fungicide that made the seed sterile so it couldn't grow.'

Greenpeace took samples of corn imported from the USA in March 1999, analysing samples from three different boats docked in Veracruz. The results showed that it was *Bt* corn made by Novartis. The campaigning group even planted some of the seeds and grew them, making sure to harvest them before they released pollen. Then they took the GM corn to the Ministry of Agriculture. 'We told them it could grow, but they said it would not happen. They have done nothing to stop or solve the problem,' says Magallon Larson. Despite Greenpeace's concerns, Dr Chapela says that: 'We were not expecting to find transgenics when we went looking for them in Oaxaca'.

Although they were working in Mexico, Chapela's and Quist's academic base is in Berkeley, where Chapela is an assistant Professor. Although a microbial ecologist by training, he had served on the prestigious National Research Council's Committee on Environmental Impacts Associated with the Commercialization of Transgenic Plants, whose report was published in 2002 by the National Academy Press.[4]

Both scientists had sprung to prominence in 1988 as two of the key opponents of a multi-million dollar alliance between Novartis and the

University of Berkeley. Unbeknown to Chapela and Quist at the time, their opposition to the Novartis deal would come back to haunt them after their research was published. The ensuing saga led to the most acrimonious fight between opponents and proponents of GM since the Pusztai affair. It also laid bare a central strategy of the biotechnology industry: that of GM contamination, and raised questions about what many believe is one of its Achilles' heels: that it could be inherently unstable. The argument over whether Quist and Chapela were attacked because they did bad science or because they questioned GM continues to run and run.

Back in the laboratory, Quist and Chapela starting using the standard amplification technique for DNA called polymerase chain reaction. Known as PCR for short, it is used to test 'for the presence of a common element in transgenic constructs' and in this case that was the promoter for the CaMV virus. The CaMV, the promoter at the heart of the Pusztai controversy, is seen as an ideal marker to tell if transgenic contamination has occurred.[5] But the PCR technique can also be problematic, as the amplification process can cause 'false positives' where simple contamination in the lab can seem to be part of the transgenic DNA. So researchers can believe they are looking at genetic contamination when in fact they are looking at experimental contamination.

Chapela and Quist also analysed control samples that came from maize grown in Peru and from seeds from the Sierra Norte de Oaxaca region in Mexico taken in 1971, long before the introduction of GM crops. They found positive PCR amplification in four of the six samples of the Oaxaca maize, but no contamination in the Peruvian maize or the older sample.[6]

They then undertook a further similar analysis, called inverse PCR, so that they could establish the precise position of the transgenic sequences. They were able to identify the DNA fragments flanking the CaMV promoter sequence through inverse PCR tests, known as iPCR. The fragments were scattered about in the genome, suggesting a random insertion of the transgenic sequence into the maize genome.[7]

So essentially, Quist and Chapela reached two conclusions. The first was that GM contamination had occurred in Mexican maize and the second was that the GM DNA seemed to be randomly fragmented in the genome of the maize. If the first point was contentious, the second was explosive, as it suggested that transgenic DNA was not stable.

Quist and Chapela knew that if the research was published it would cause an international outcry, so they wanted to make sure that their research was correct. The biotech industry had hardly recovered from the StarLink scandal in the USA, and GM contamination of Mexican maize would represent a 'nightmare' scenario for the industry.[8]

'I repeated the tests at least three times to make sure I wasn't getting false-positives', says Quist.[9] Convinced of their findings, Chapela shared the preliminary results with various Mexican government officials who started to do their own testing. He also approached the scientific journal *Nature* with a view to publishing the work.

'I had been talking to government officials, because I thought it was the responsible thing to do, even though it was preliminary research', recalls Dr Chapela.[10] At one meeting the aide to the Biosafety Commissioner, Fernando Ortiz Monasterio, told Chapela that his boss wanted to see him. 'The guy just sat outside the door and when I came out, he almost took me by the hand and put me in a taxi with him to see his boss,' he says.

A Hollywood script-writer could have conceived what happened next. Chapela was hauled up to Monasterio's 'office' on the 12th floor of an empty building. 'The office space was absolutely empty', recalls Chapela. 'There were no computers, no phones, the door was off its hinges, there were cardboard boxes as a table. The official is there with his cell-phone beside him. We are alone in the building. His aide was sitting next to me, blocking the door.'

With obvious emotion, Dr Chapela recalls what happened next. 'He spent an hour railing against me and saying that I was creating a really serious problem, that I was going to pay for. The development of transgenic crops was something that was going to happen in Mexico and elsewhere. He said something like I'm very happy it's going to happen, and there is only one hurdle and that hurdle is you.'

Sitting stunned, Chapela replied: 'So you are going to take a revolver out now and kill me or something, what is going on?' Then Monasterio offered Chapela a deal: 'After he told me how I had created the problem, he said I could be part of the solution, just like in a typical gangster movie. He proceeded to invite me to be part of a secret scientific team that was going to show the world what the reality of GM was all about. He said it was going to be made up of the best scientists in the world and you are going to be one of them, and we are going to meet in a secret place in Baja, California. And I said, "who are the other scientists"', and he said "Oh I have them already lined up, there are two from Monsanto and two from DuPont". And I kept saying "Well that is not the way I work, and I wasn't the problem, and the problem is out there".'

Then events took a very sinister turn. 'He brings up my family', recalls Chapela. 'He makes reference to him knowing my family and ways in which he can access my family. It was very cheap. I was scared. I felt intimidated and I felt threatened for sure. Whether he meant it I don't know, but it was very nasty to the point that I felt "why should I be here, listening to all this and I should leave".'

Monasterio later admitted to the BBC that he had met Chapela, but vehemently denied threatening him in any way. He said that the meeting had taken place not on the 12th floor, but on the '5th floor of our offices, which is an office of the Ministry of Health, in the southern part of town where we work'. He said that at the meeting they had discussed 'the issues of the presence of maize, the importance of publishing, that what we were doing is research, and that when we have the results from our own researchers, we will share with him'.[11]

Chapela was told by Monasterio that he was in charge of biosecurity and 'I'll tell you what biosecurity is really about, it is about securing the investment of people who have put their precious dollars into securing this technologies, so my job is to secure their investment'.

'I think first he was trying to intimidate me into not publishing,' says Chapela. Once Monasterio realized that Chapela was going to try and publish his results, that 'very night he called a meeting with Greenpeace and the people from Codex and people from the Senate to divulge the results'.

The reason that Monasterio wanted the results made public was simple: 'I had said to him', says Chapela, 'that if the information was released before it was published in *Nature* then *Nature* would think twice about publishing it'. 'He fed it directly to Greenpeace, which is a lot easier to discredit than *Nature*,' says Chapela, adding that Monasterio knew that 'the media coverage would seriously threaten publication in *Nature*'. Monasterio denies breaking any confidentiality agreement by divulging the results early.[12]

But the threats intensified against Chapela, who received a letter from an agricultural under-secretary, saying that the government had 'serious concerns' about the 'consequences that could be unleashed' from his research. Moreover the government, would 'take the measures it deems necessary to recuperate any damages to agriculture or the economy in general that this publication's content could cause'.[13]

'He signed it before the publication is out and it is obvious that he is trying to intimidate me into not publishing', says Chapela, who believes that the approach is not surprising, as the Agriculture Ministry itself is 'riddled with conflicts of interest. There are just working as spokespeople for DuPont, Syngenta and Monsanto'.

In contrast to the agricultural officials, others were worried, and started to replicate the research. As Quist and Chapela outline: 'During the review period of this manuscript, the Mexican government ... established an independent research effort. Their results, published through official government press releases, confirm the presence of transgenic DNA in landrace genomes in two Mexican states, including Oaxaca'.[14] On 17 September 2001, Mexico's Secretary for Environmental and Natural Resources released partial results of its own study, confirming that transgenic maize had been found in 15 of 22 areas tested in Oaxaca and nearby Puebla.[15]

Just over two months later, Chapela's team published in *Nature*. 'We report', wrote Chapela and Quist, 'the presence of introgressed transgenic DNA constructs in native maize landraces grown in remote mountains in Oaxaca, Mexico, part of the Mesoamerican centre of origin and diversification of this crop'. In plain English, they were reporting contamination of native corn by its GM equivalent.

The scientists were both 'surprised and dismayed' over their findings, but admitted they had no way of knowing whether the contamination was from a loose implementation of the moratorium or due 'to introgression before 1998 followed by the survival of transgenes in the population'.[16]

'Whatever the source, it's clear that genes are somehow moving from bioengineered corn to native corn', says Chapela. 'This is very serious because the regions where our samples were taken are known for their diverse varieties of native corn, which is something that absolutely needs to be protected. This native corn is also less vulnerable to disease, pest outbreaks and climatic changes.'[17]

Once again it was time to shoot the messenger. 'We are just facing every single level of intimidation and aggression that you can imagine', says Chapela. 'It is obviously very well funded and very well coordinated'.

'The main attack, the most damaging attack' came 'from my own colleagues within the university', says Chapela, 'who are mad at me because I stood up against Novartis coming in with US$50 million and buying the whole college. It has to be said that the immediate consequences might be very dire for me as my tenure is being reviewed.' Chapela says that because of his stand against Novartis: 'They are saying that we are activists, that we are anti-biotech'. Ironically before joining the staff at Berkeley Chapela had worked for Sandoz, which later merged with Ciba-Geigy to form Novartis.[18]

Some of the most virulent attacks came via the AgBioView discussion group and AgBioWorld.org website run by C S Prakash, who is a Professor of Plant Molecular Genetics at Tuskegee University, Alabama. Prakash's foundation and website are an influential talking shop for GM scientists world-wide and a key place to influence other scientists. But the underlying reason for its existence is the promotion of biotechnology and the website features a Declaration in Support of Agricultural Biotechnology, signed by more than 3300 scientists from around the world, including 19 Nobel Prize winners.

Prakash calls the Quist and Chapela study 'flawed', saying that the 'results did not justify the conclusions'. He says that they were 'too eager to publish their results because it fitted their agenda'. A co-founder of the AgBioWorld Foundation, is Gregory Conko, from the right-wing free enterprise think-tank the Competitive Enterprise Institute (CEI), based in Washington. The CEI has a long history of working with the anti-environmental 'Wise Use' movement, and is a key player in the backlash against people speaking out on environmental issues.[19]

Prakash says that the AgBioWorld website 'played a fairly important role in putting public pressure on *Nature*, because we have close to 3700 people on AgBioView, our daily newsletter, and immediately after this paper was published, many scientists started posting some preliminary analysis that they were doing'.

'It was not just the paper from Chapela that was damaging from the point of view of biotechnology', says Prakash. 'But a large number of media interviews, where he claimed that Mexican biodiversity was contaminated, the ability to feed its people was threatened, really outlandish claims that probably irked many of the scientists.'

The first attack came on Prakash's website within hours. But it was not a scientist who fuelled the attacks, but someone called Mary Murphy. 'The activists will certainly run wild with news that Mexican corn has been "contaminated" by genes from GM corn not currently available in Mexico... It should also be noted that the author of the *Nature* article, Ignacio H Chapela, is on the Board of Directors of the Pesticide Action Network North America (PANNA), an activist group' wrote Murphy. Chapela was 'not exactly what you'd call an unbiased writer'.[20]

The next AgBioView bulletin led with a posting from someone called Andura Smetacek, under the head-line 'Ignatio Chapela – activist FIRST, scientist second'. It read: 'Chapela, while a scientist of one sort, is clearly first and foremost an activist'. 'Searching among the discussion groups of the hard-core anti-globalization and anti-technology activists Chapela's references and missives are but a mouse click away.'

Smetacek argued that the article was 'not a peer-reviewed research article subject to independent scientific analysis'. Her email included detailed information on the author and tried to undermine his credibility. 'A good question to ask of Chapela would be how many weeks or months in advance did he begin to coordinate the release of his "report" with these fear-mongering activists [Greenpeace, Friends of the Earth]? Or more likely, did he start earlier and work with them to design his research for this effect?'[21] In the space of an email, peer-reviewed research becomes non peer-reviewed research designed by 'hard-core' environmental groups.

In the next bulletin, on 30 November, other contributors continued the theme started by Smetacek and Murphy, questioning Quist and Chapela's 'activist' links and their research. 'Mary Murphy's comment echoes my reaction when I read the news reports... This alarmist reporting of preliminary, incomplete research is just another example of the nutty illogic of the anti-GE luddites.'[22]

These attacks by Smetacek and Murphy were sent immediately after the publication of the *Nature* article. Their character assassinations set the tone for others to follow, as we shall see. They had moved the debate from the message to the messengers and it was time for character assassination. Even the journal *Science* noted the part played by what it called, 'widely circulating anonymous emails' accusing researchers, Ignacio Chapela and David Quist, of 'conflicts of interest and other misdeeds'.[23] Some scientists though, were alarmed at the personal nature of the attacks. 'To attack a piece of work by attacking the integrity of the workers is a tactic not usually used by scientists', wrote one.[24]

A virtual world

So who are Mary Murphy and Andura Smetacek, who between them have posted over 60 articles on the Prakash site? Mary Murphy's email is mmrph@hotmail.com, which seems like just another hotmail address. However, on one occasion, Murphy posted a fake article claiming that Greenpeace had changed its stance on GM due to extra strength GM marijuana. Although Murphy used her hotmail address mmrph@hotmail.com, she left other identifying details, including 'bw6.bivwood.com'.[25]

Bivwood is the email address for Bivings Woodall, known as the Bivings Corporation, a PR company with offices in Washington, Brussels, Chicago and Tokyo. Bivings has developed 'internet advocacy' campaigns for corporate America[26] and has been assisting Monsanto on internet PR ever since the biotech company identified, in 1999, that the net had played a significant part in its PR problems in Europe. While Bivings claims its work for Monsanto is an example of how it approaches contentious issues in a 'calm and rational way', it uses the internet's 'powerful message delivery tools' for 'viral dissemination'.

As it outlines: 'Message boards, chat rooms, and listservs are a great way to anonymously monitor what is being said. Once you are plugged into this world, it is possible to make postings to these outlets that present your position as an uninvolved third party.'[27]

But evidence points to the fact that Bivings, or those who have had access to its email accounts, has covertly smeared biotech industry critics via a website called CFFAR (Center for Food and Agricultural Research), although no such organization appears to exist, as well as articles and attacks posted to listservs under aliases. The attack on the *Nature* piece is a continuation of this covert campaign.[28]

Andura Smetacek is the original source of a letter that was published in *The Herald* newspaper in Scotland under the name of Professor Tony Trewavas, a pro-GM scientist from the University of Edinburgh. This letter was the subject of a legal action between Greenpeace, its former director Peter Melchett and the newspaper. The case went to the High Court and resulted in Peter Melchett being paid damages, which he donated to various environmental groups, and an apology from *The Herald*.[29]

Trewavas has always denied that he wrote the defamatory letter, and Andura Smetacek has acknowledged that 'I am the author of the message, which was sent to AgBioWorld. I'm surprised at the stir it has caused, since the basis for the content of the letter comes from publicly available news articles and research easily found on-line'.[30]

Despite the email address, Andura Smetacek is also a 'front email'. Although in early postings to the AgBioView list, she listed her address as London, in a dispute with *The Ecologist* magazine Andura left a New York phone number. However, enquiries have discovered that there is no person of that name on the electoral roll or other public records in the USA. Despite numerous requests to give an employer or verify a land address for *The Ecologist*, Smetacek has refused to do so.[31] Subsequent attempts by both British and American journalists to track down Smetacek have also elicited no answers.[32]

The first exposé of the Bivings connection to the *Nature* article was published by myself in the *Big Issue* magazine, and by the anti-GM campaigner, Jonathan Matthews, in *The Ecologist* magazine.[33] Bivings denied being involved in the dirty tricks campaign, saying that the reports were 'baseless' and 'false', and merited 'no further discussion'.[34]

Environmental commentator George Monbiot subsequently published two articles in *The Guardian*. 'The allegations made against the Bivings Group in two recent columns are completely untrue,' responded Gary Bivings, President of the Group. Bivings also contended that 'the 'fake persuaders' mentioned in the articles – Mary Murphy and Andura Smetacek – are not employees or contractors or aliases of employees or contractors of the Bivings Group. In fact, the Bivings Group has no knowledge of either Mary Murphy or Andura Smetacek'.[35]

BBC *Newsnight* then took up the story. A spokesperson for Bivings admitted to a researcher from *Newsnight* that 'one email did come from someone "working for Bivings" or "clients using our services"'. But once again they denied an orchestrated covert campaign.[36] Bivings later argued that they had 'never made any statements to this effect', saying that BBC *Newsnight* had been 'wrong'.[37]

Gary Bivings also denied any involvement with the CFFAR website. But the website is registered to an employee of Bivings, who was a Monsanto web-guru. Furthermore, Bivings denied any involvement with the AgBioWorld Foundation, yet Jonathan Matthews had received an error message whilst searching the AgBioWorld database that a connection to the Bivings server had failed. Internet experts believed that this message implied that Bivings was hosting the AgBioView database. These experts also noticed technical similarities between the CFFAR, Bivings and AgBioWorld databases.[38]

Prakash, however, denied receiving funding or assistance for the AgBioWorld Foundation and denied working with any PR company, saying that he is 'pro-the technology not necessarily the companies'.

There were other tell-tell signs to be found, too. For people who had been so prolific in their attack against Chapela, Mary Murphy and Andura Smetacek suddenly disappeared. Murphy's last posting was on 8 April, just a few days before the *Big Issue* piece went out. That same month, April 2002, Bivings had posted an article by their contributing editor, Andrew Dimmock, called 'Viral Marketing: How to Infect the World' on the web.

However, after the story broke in the UK press, Bivings changed their on-line version. Out went the sentence 'There are some campaigns where it would be undesirable or even disastrous to let the audience know that your organization is directly involved' and out went the 'anonymously'. One sentence was changed from 'present your position as an uninvolved third party' to 'openly present your identity and position'.[40] In the autumn of 2002, Bivings outlined how the term viral marketing had been 'unfairly vilified' in the press, it was nothing more than 'word-of-mouth advertising via the internet'.[41]

Why would a company that had nothing to do with the *Nature* attack, suddenly change articles on its website? Even more intriguing were the actions of a Biving's web designer who lived in the locality of the server that had posted the 'Mary Murphy' emails. Having worked at Bivings for seven years, as a senior programmer, this person suddenly changed his on-line CV, deleting all references to Bivings. Suddenly he had spent the last seven years being a 'Freelance Programmer/Consultant'. The only problem is that his old CV is still on-line in an archive site that repeatedly mentions that he had worked for Bivings.

There was one other slight but important change to the Bivings site that occurred after the publicity too. Bivings had listed 15 different Monsanto websites as clients, however this changed to just a direct link to Monsanto.com afterwards. Were Bivings trying to hide just how much work they did for Monsanto? Once again, you can see the old version on internet archive sites.[42] Finally, the CFFAR website was suspended, with the site hosting an inoffensive 'holding page', but once again it is still available on archive sites.[43]

Monsanto denied that it employed Bivings to undertake this kind of work. 'They don't do PR', said a Monsanto spokesperson. 'We speak for ourselves on issues'.[44] This begs the question of what kind of work Monsanto do on the web, and finally solves the mystery of the identity of Andura Smetacek. The company has radically changed its on-line activities in the last few years. After Monsanto's European PR took a 'beating' in 1999, Monsanto's communications director said 'maybe we weren't aggressive enough. When you fight a forest fire, sometimes you have to light another fire'.[45]

In January 2000, Prakash had set up the AgBioWorld website.[46] In July 2000, Andura Smetacek suddenly appeared on AgBioView, writing in a very measured tone. 'While I remain concerned about who controls biotechnology', wrote Smetacek. 'I have come to a disturbing conclusion about some of the groups with whom I have been discussing this issue who so strongly oppose genetic engineering. Their tactics and support for violence and vandalism are unacceptable and must stop.'

Smetacek then mentioned the recently registered CFFAR site, saying that she had 'signed a petition to stop these acts of terrorism posted to www.CFFAR.org'. At the time Smetacek gave a London address, although the time and date on the email located it as 'Pacific Day Time', coming from the Pacific Coast of the USA.[47] In the first months of the AgBioView list, messages were forwarded in such a way that it was possible to track the technical 'headers' that shows where a message comes from. The first few from Andura showed they had come from '199.89.234.124'. If you look up these numbers they are assigned to Monsanto in St Louis, Missouri. So, from the email address, it seems that Andura Smetacek writing from London never actually existed, 'she' was a virtual person whose role was to direct debates on the web and denigrate the opposition.

When asked what work they did for Monsanto, a spokesperson for Bivings said 'We run their websites for various European countries and their main corporate site and we help them with campaigns as a consultant and we are not allowed to discuss strategy issues and personal opinions'. They declined to give further details of their work for the biotech company,[48] but they suggested talking to another PR company that worked for Monsanto, called V-fluence.

The contact person given was Rich Levine, who previously worked for Bivings as a Monsanto web-guru.[49] The president of V-fluence is Jay Byrne, who has over 15 years experience in public relations, campaign communications and government affairs.[50] He was also the former chief internet strategist and director of corporate communications for

Monsanto, where he spent a quarter of his time monitoring the web for rogue web- and activist sites.[51]

In 2001, Byrne gave a presentation to a PR conference called 'Protecting Your Assets: An Inside Look at the Perils and Power of the Internet'. It gave an insight into Monsanto's use of the internet. 'A website alone won't protect your brand', Byrne told the audience, therefore it was necessary to 'Take Action, Take Control'. Ways to do this included: 'Viral marketing and other dialogue opportunities, monitoring and participation'.

One PowerPoint slide showed 'Monitoring' for Monsanto which included 'Daily monitoring of over 500 competitor, industry, "issues group" websites; Daily monitoring of 50+ key listservs, usergroups and chat rooms; Technology monitoring and updates including search engine programs and legal monitoring'.

Another chart on the PowerPoint presentation gave the difference before and after taking control of the internet to rig a search engine to go from finding hits they did not want to finding hits they did want if someone was searching for 'GM food'. Favourable hits included: 'Glossary of biotech terms; AgBioWorld; AgCare; FDA; Biotech Knowledge Center; CFFAR; Food Biotech Center; and Biotech Basics'. To the uninitiated these would all appear as independent sites, yet we now know that three of these are acknowledged Bivings projects – BioTech Terms; Biotech Knowledge Center and Biotech Basics. Two seem to have links to Bivings – AgBioWorld and CFFAR. One – AgCare – is a biotech lobby front in Canada, and the other – the US FDA – is seen by the biotech industry as an ally.

Of these, the CFFAR site is the most worrying in that it denigrates environmentalists as terrorists. It is the site that Andura wanted the scientists to look at. Once you denigrate someone it becomes easier to attack them, both physically and mentally and even intellectually. Byrne finished by quoting Michael Dell, the CEO of Dell computers: 'Think of the internet as a weapon on the table. Either you pick it up or your competitor does – but somebody is going to get killed.'[52]

The fall-out continues

In January 2002, the Mexican Ministry of the Environment confirmed their findings from the previous year and said that in some remote regions of Oaxaca and Puebla, between 20–60 per cent of tested farms had traces of transgenic material.[53]

The following month Chapela appeared at a press conference with Mexican researchers. Chapela had given some samples to the

Environment Ministry who had divided the samples. One batch had been sent to the National University and the other to the Centre for Investigation and Advanced Studies. Both gave details of preliminary research that backed Chapela's findings.[54]

'They have reworked that study in two separate labs, with new sampling and new methodology. Last week, when I was in Mexico', he says when interviewed in March 2002, 'they were announcing that they were close to publication and that everything they had pointed in the same direction and they supported our work. Their principal investigator says they have three levels of analysis – the DNA, the protein and the expression level of analysis and everything that I have seen so far makes it extremely unlikely that there are any mistakes in our statement to *Nature*.'

So Chapela says that there are now three separate studies that have been done by two separate groups that 'confirm what we are saying, down to the quantitative level. I am still hopeful that I am not going to end the way Pusztai has seen himself pushed out of his job and discredited for publication in major journals. I think and I hope that we will be vindicated'.

But despite his optimism, in February 2002, the row intensified when an editorial written by Paul Christou, then at the John Innes Centre, appeared in the journal *Transgenic Research*. It was brutal. Its title said it all: ' No Credible Evidence is Presented to Support Claims that Transgenic DNA was Introgressed into Traditional Maize Landraces in Oaxaca, Mexico'.

Christou, writing on behalf of the Editorial Board, wrote that Quist and Chapela's paper had 'technical and fundamental flaws'. Sample contamination was the likely cause of the results, not GM contamination. This said, Christou pointed out that 'introgression of transgenes from commercial hybrids into landraces is likely'.[55]

'Recombination is not a satisfactory explanation either, since multiple generations of crossing have been done with all these constructs, and they have been shown to be stable – or else they would have not made it through the regulatory system,' wrote Christou. Critics of the industry say that whilst Christou's statement is broadly correct, the applicable regulatory standard for a demonstration of 'stability' is low, especially in the USA.[56]

Moreover, critics of the biotech industry point to regulatory laxness again. Consider the EPA's analysis for the stability of *Bt* crops. In its re-registration document for *Bt* crops in 2001, the EPA noted that 'stability and inheritance were not addressed with the registrations' for Monsanto's *Bt* corn and potato. The EPA said that because these crops had been growing for a number of years with a lack of reports relating to loss of efficacy, 'this specific endpoint can be considered to have been addressed through commercial use'.[57]

So because the EPA has not been notified of any failures, the products are deemed to be 'stable'. This is exactly the same unscientific analysis whereby, because the authorities have not been notified of any ill effects, GM products are deemed to be 'safe'.

Chapela called the *Transgenic Research* article a 'regurgitation' of old arguments, but it angered others working on the issue. Peter Rosset from Food First, a think-tank, called it 'a "hit piece" designed to leave the public with a sense of confusion about whether the contamination was real or not'. He continued, citing Pusztai as an example that: 'I firmly believe there is a concerted attempt to make "examples" of scientists who have the courage to be dissidents from the biotech juggernaut. Clearly industry – and scientists on the industry gravy train – want to stifle scientific dissent, and cast a smoke screen over the public's perception of the risks of GMOs'.[58]

Scientists working in the field agree. Sue Mayer from GeneWatch UK says that 'it is quite extraordinary the lengths the biotech industry and scientific establishment will go to discredit any critical science'.[59] Professor Allan McHughen, from the Crop Development Center at the University of Saskatchewan in Canada, believes that there 'are a group of people who for whatever reason don't want to hear anything at all about reasons to question the technology. I read Chapela's paper over and over again and I just couldn't find anything that was inflammatory about it'.[60]

'I don't think the science in the second half of their paper was very good,' adds Allison Snow of Ohio State University, who specializes in gene flow. 'But the first half of the paper, while you could always have asked them to do a better job, I thought was well supported. The things they said could have been taken as a threat to the field of ag biotechnology because all along the ag biotechnologists have been saying that we know what these genes do, they're just like other genes.'[61]

Statements for and against

However, if the industry thought that threatening and undermining Chapela would make the controversy disappear, they were wrong. One of the leading anti-GM protagonists in the USA is Ronnie Cummins of the Organic Consumers Association. 'What the biotech industry is underestimating', says Cummins is that, 'corn is not just another crop down here. It is central to the culture. It is a total insult to the people in Mexico as to what is going on.'

The Organic Consumers Association and Food First were two of the 144 farmer and other civil society organizations from 40 countries that signed a statement on the Mexican GM Maize scandal in February 2002. It stated that 'A huge controversy has erupted over evidence that the Mesoamerican Center of Genetic Diversity is contaminated with genetically modified maize. Two respected scientists are under global attack and the peer-review process of a major scientific publication is being threatened'. The signatories claimed that 'pro-industry academics are engaging in a highly unethical and mud-slinging campaign against the Berkeley researchers'.[62]

On the AgBioView list, this document provoked outrage and the attacks against Chapela intensified. Alex Avery is a well-known adversary of organic food (see Chapter 10). Alex works with his father, Dennis, at the Centre for Global Food Studies that is affiliated to the right-wing think tank, The Hudson Institute. 'Has anyone else picked up on the "Joint Statement on the Mexican GM Maize Scandal" being whored around by the anti-biotech activists?' asked Alex Avery.

Avery followed Smetacek's and Murphy's lead. 'Chapela is an activist assistant professor of microbiology… He isn't a geneticist, but he is on the board of Pesticide Action Network North America (an anti-pesticide activist group) and in 1999 signed an anti-biotech statement calling for a global moratorium on GM crops'. Avery then said that Chapela and Quist were 'far from the "respected scientists" that the Joint Statement claims. 'Then again', wrote Avery 'they do their darndest to paint Arpad Pusztai as a "widely respected scientist" in the statement, despite the drubbing Pusztai's research and methodology took from The Royal Society experts.'

Avery then proposed that 'Fellow scientists, perhaps we should get out front on this and post a "joint statement" from academics.'[63] In a statement posted on AgBioWorld.org on 24 February 2002, Prakash wrote that 'the research methodology and its conclusions are however being challenged by a number of groups through formal letters to *Nature* (under review), and it was also addressed recently in an editorial in the Journal 'Transgenic Research'. He urged subscribers to the list to sign the petition.[64]

When is a retraction not a retraction?

Finally on 4 April 2002, *Nature* issued a terse statement on its website that there was disagreement between the Quist and Chapela and one reviewer. Because of this and 'several criticisms of the paper *Nature* has concluded that the evidence available is not sufficient to justify the publication of the original paper.'[65]

'It is clearly a topic of hot interest,' said Jo Webber from *Nature*, admitting the story was not just 'technical' but also 'political'. '*Nature* has been going for a very long time and this is a very unusual occurrence'. Webber also admitted that she felt her editor had fudged the issue.[66]

The statements put out by *Nature* seemed to be contradictory and there was confusion as to whether the paper had actually been 'retracted'. The Editor, Philip Campbell, wrote 'The retraction was necessitated by technical flaws in the paper that came to our attention after its publication (which we should have picked up), and by the authors' decision not to retract the paper themselves'.[67]

In contrast, Dr Maxine Clarke, the Executive Editor of *Nature* wrote a month later in June that the Quist and Chapela paper 'has not been formally retracted by *Nature*, and stands as a citable publication'.[68] Quist certainly felt it was a fudge: 'I think they wrote in very specific language for a reason, so that it was somewhat equivocal', he says. 'If results come out to corroborate our results, they can say, "See, we didn't ask for a retraction because it is a biological reality; it is happening". If it turns sour, they can say, "See, we were right in putting these guys on the chopping block".'[69]

Chapela was more blunt, accusing Campbell of 'siding with a vociferous minority in obfuscating the reality of the contamination of one of the world's main food crops with transgenic DNA of industrial origin'.[70] Campbell had sent the paper to three referees before deciding whether to retract. Of the three, only one scientist thought the paper should be retracted – though all said there were flaws in its second part – the section on iPCR. Others joined in the argument, and the journal was accused of setting a 'dangerous precedent' and it was added that, 'by taking sides in such unambiguous manner, *Nature* risks losing its impartial and professional status'.[71]

Due to the connections between the prominent attackers and the biotech industry, Chapela requested that *Nature* print a 'statement of conflict of interest from all authors,' as regarding the Berkeley–Novartis connection. 'It cannot go unnoticed that the antagonists signing the letter against the *Nature* piece should all be connected directly with this local political scandal', wrote Chapela. Campbell refused.

Chapela also noted that 'Given that two of the three reviewers of the exchange between our critics and ourselves unequivocally state that our main results and statements are not legitimately challenged by the letters included here, we find it unjustified that *Nature* should decide to remove its endorsement of a paper which itself was subjected to several rounds of a particularly stringent review process'.

Chapela noted how the second referee had said 'none of the critics seriously dispute the main conclusion' and the third said, 'none of the comments has successfully disproven their main result that transgenic corn is growing in Mexico and crossing with local varieties'. Yet Dr Campbell published the retraction – citing only the first referee, leading to the charge that 'he had ignored the advice of most of its own advisers'.[72]

In the end *Nature* published two critical letters, one from a team led by Nick Kaplinsky in the Department of Plant and Microbial Biology – the department at Berkeley that received the Novartis funding. The lead author of the other letter was Matthew Metz, who also used to be at the Department of Plant and Microbial Biology at Berkeley.[73]

Both lead authors – Matthew Metz and Nick Kaplinsky – were signatories to the Prakash 'Joint Statement' that Prakash had urged scientists to sign. It has received nearly 100 signatories.[74] Metz had co-edited a pro-biotech document with the AgBioWorld Foundation, the Liberty Institute and the Competitive Enterprise Institute two years before.[75] Another co-editor was Andrew Apel, editor of the industry newsletter, *AgBiotech Reporter*, who used the 11 September attacks to vilify anti-GM activists and scientists, specifically Drs Vandana Shiva and Mae-Wan Ho, as having 'blood on their hands'.[76]

In his letter to *Nature*, Metz argued that Quist and Chapela's analysis was 'flawed' and that the authors had 'misinterpreted' a key reference. Kaplinsky's letter argued that Quist and Chapela may have been 'confused', and although transgenic corn could be growing in Mexico, their claims were 'unfounded'.[77]

Chapela admits that *Nature* was 'under incredible pressure from the powers that be', and that the journal had asked him to respond to four letters that were critical of his paper, of which only the Kaplinsky and Metz letters were published. Both of these critics work or used to work at the department that received the Novartis funding. Metz's co-author, Johannes Fütterer, is a post doctorate student at ETH-Zurich, under Wilhelm Gruissem. According to Chapela 'Gruissem was head of department in Berkeley and the person who brought Novartis to us'.

Chapela believes that it is this issue that lies at the heart of the whole saga. 'I and a few other people stood up against it and we made a big scandal that went around the world. It became a very big scandal', he says. 'And they just cannot forgive that.' Metz had even written to *Nature* defending the Novartis deal.[78] Chapela points to an article in the German press that says that Fütterer only 'decided' to write the letter with Metz after consultation with his boss, Gruissem, and 'his American research associates'.[79] So everyone who had letters published in *Nature* was in some way connected to the Novartis-Berkeley relationship.[80]

This point was also taken up by others, pointing out the controversy was taking place 'within webs of political and financial influence that compromise the objectivity of their critics'. Correspondence to *Nature* also pointed out that the 'Nature Publishing Group actively integrates its interests with those of companies invested in agricultural and other biotechnology, such as Novartis, AstraZeneca and other "sponsorship clients", soliciting them to "promote their corporate image by aligning their brand with the highly respected *Nature* brand"'.[81] As if to prove their point, just over six weeks later, *Nature* ran a special 'Insight' into food and the future, sponsored by Syngenta that contained several pro-GM and anti-organic-farming opinion pieces.[82] But Metz and Kaplinsky replied that their criticisms of Quist and Chapela, were 'exclusively over the quality of the scientific data and conclusions' and that their funding has 'absolutely nothing to do' with their criticisms.[83]

However, the journal also published a further letter by Quist and Chapela where they acknowledged that in relation to iPCR they had misidentified certain sequences. But they added 'the consistent performance of our controls, as reported, discounts beyond reasonable doubt the possibility of false positives in our results'. The authors, noted that 'to address' the challenges laid down by their critics they had used a 'non-PCR-based method' called DNA–DNA hybridization. 'The results of these experiments' they argued, 'continue to support our primary statement... The DNA-hybridization study confirms our original detection of transgenic DNA integrated into the genomes of local landraces in Oaxaca.'[84]

Ironically the fact that GM contamination has occurred is now not disputed by the GM opponents. 'Quist and Chapela have subsequently presented data that further supports the presence of transgenes in maize landraces – a point that has not been disputed', argued Prakash on AgBioWorld.[85]

In April, Jorge Soberon, the executive secretary of Mexico's National Commission on Biodiversity, announced the findings of the Mexican government's research at the International Conference on Biodiveristy at The Hague. Soberon confirmed that the tests had now shown the level of contamination was far worse than initially reported in both Oaxaca and Puebla. A total of 1876 seedlings had been taken by government researchers and evidence of contamination had been found at 95 per cent of the sites. One field had 35 per cent contamination of plants alone. The Mexican government also re-confirmed the presence of the Cauliflower Mosaic Virus.[86]

Jorge Soberon said soberly that: 'This is the world's worst case of contamination by genetically modified material because it happened in the place of origin of a major crop. It is confirmed. There is no doubt about it'. In response, Philip Campbell, the editor of *Nature*, said: 'The Chapela results remain to be confirmed. If the Mexican government has confirmed them, so be it'.[87]

In August the President of Mexico's National Institute of Ecology, confirmed that his team had found 7 per cent of the native maize plants they sampled contained genetic material that appeared to come from bioengineered corn. 'This is basically the same result that Chapela reported in his study, and both results suggested the presence of transgenic constructs in native maize varieties', he said, confirming that the paper had been submitted for publication.[88]

But two months later, the controversy took a new twist when the Mexican press announced that *Nature* had rejected their independent studies into the GM contamination for publication. The reviewers had rejected the papers for opposing reasons. One said that the results were so 'obvious' that they did not merit publication in a scientific journal, whereas the other said the results were 'so unexpected as to not be believable'. The *Nature* editor said the papers had been rejected on 'technical grounds'.[89]

So over a year after the revelation of GM contamination in Mexico, the controversy continues and nothing has been done to stop the source of the contamination, but then perhaps that is what the industry wants.

Is GM contamination beneficial?

In the Joint Statement signed by Kaplinsky, Metz and Prakash there is one paragraph that stands out as warranting further analysis: 'It is important to recognize that the kind of gene flow alleged in the *Nature* paper is both inevitable and welcome.'[90]

So GM contamination is not only inevitable but also beneficial, and it fuses together two important pro-biotech messages: that biotechnology is no more than an extension of traditional plant breeding and that because contamination is inevitable, any kind of resistance is futile. Contamination could be inevitable unless regulators act. As *Nature Biotechnology* candidly pointed out, 'gene containment is next to impossible with the current generation of GM crops ... gene flow from GM crops to related plants thus remain a primary concern for regulators and one that companies need to address'.[91]

Ironically it is in the biotech companies interests not to address this problem, although that is not in the interests of consumers who want choice. 'The hope of the industry is that over time the market is so flooded [with GMOs] that there's nothing you can do about it, you just sort of surrender', say Don Westfall, vice-president of Promar International, a consultant to the biotech and food industries in Washington.[92]

Critics of the biotech industry cannot believe what they read in the Prakash statement. 'It is not beneficial for the Mexican campesinos or peasants or indigenous peoples', says Hector Magallon Larson, from Greenpeace Mexico. 'It is not beneficial for the Mexican environment and it not beneficial for world food security.'

'You would never say that BSE was inevitable or welcome,' adds Alan Simpson MP, a leading critic of the industry. 'The arrogance of it is outstanding. One of the things that Pusztai has been trying to get us to understand is what we are talking about is a completely new frontier and it's not about plant breeding. This is being run past society and past political institutions on the basis that it is both a radical scientific advance and yet no different at all. It is unbelievably dishonest and anti-scientific.'

There are numerous reasons why the process cannot be beneficial, and one of these is the potential inherent instability of GM crops, something that was outlined in the discussion of the Pusztai saga in Chapter 5 and which Quist and Chapela still stand by. 'It suggests that transgenic DNA can move around the genome with a range of unpredictable effects, from disruption of normal functions to modification of expressed products that become toxic agents to the generation of new strains of bacteria and viruses,' Quist says.[93]

'There are a lot of theoretical reasons to believe that most of the transformation events are going to be ultimately unstable, particularly as they have been put in another environment', adds GM specialist Dr Michael Hansen from the US Consumers Union.

The fact that many biotech scientists have signed on to a statement that says that GM contamination is inevitable, underpins the theory that many of the industry's critics and analysts have felt for some time. They believe that the industry has deliberately set out to contaminate both non-GM and organic crops with the implicit or explicit intention of making contamination inevitable. All hope of another alternative agriculture system simply vanishes and once that vanishes, the anti-GM fight becomes hopeless.

'I think the industry now recognise that hopelessness is their best hope', adds Alan Simpson. 'They have manifestly failed to convince the public of either the desirability or safety of GM products. Having failed to convince, having failed to co-opt or to buy the public support, they are left with coercion. Coercion comes in two forms. One is putting an arm lock over the farmers and the other is putting a choice lock on consumers.'

But it is not just the critics who argue that contamination is a deliberate policy. Dan McGuire, Program Director to the 2002 Annual Convention of the American Corn Growers Association: 'I believe that the biotech companies that market GMO seed would like to see the grain marketing system totally taken over and "contaminated" by GMOs. I expect they would see that as ending their problem'.[94]

With widespread GM commercialization, GM contamination is inevitable. There have now been episodes of GM contamination in Argentina, Austria, Bolivia, Brazil, Canada, Colombia, France, Germany, Greece, Holland, India, Japan, Korea, Luxembourg, Mexico, New Zealand, Sweden, Thailand, the UK and the USA, amongst others.[95] The health and environmental impact of these contamination episodes is unknown. But waiting in the wings are the second-generation crops, those with health and nutritional benefits, and third generation crops – with industrial, or pharmaceutical properties, known as pharm crops. These include vaccines, growth hormones, clotting agents, industrial enzymes, human antibodies, contraceptives and abortion-inducing drugs.[96]

Scientists believe that work needs to be done to stop pharm crops – which are already being grown – from contaminating other crops. If these are not contained, the US National Academy of Scientists warn that 'it is possible that crops transformed to produce pharmaceutical or other industrial compounds might mate with plantations grown for human consumption, with the unanticipated result of novel chemicals in the human food supply'.[97]

Dr Norman Ellstrand, a professor of genetics at the University of California, Riverside, and a leading expert on corn genetics, says that 'if just 1 percent of [American] experimental pollen escaped into Mexico, that means those landraces could potentially be making medicines or industrial chemicals or things that are not so good for people to eat. Right now, we just don't know what's in there'.[98]

Others are worried too. 'Most people are assuming that plants being used for these purposes [bio-pharming] will not enter the food supply, but if you assume that you need to have controls in place to make sure that does not happen,' says Michael Taylor, who used to work for the FDA and Monsanto. Some are more blunt: 'Just one mistake by a biotech company and we'll be eating other people's prescription drugs in our corn flakes', argues Larry Bohlen, from Friends of the Earth in the USA.[99]

It is not clear yet who will bear the ultimate responsibility for GM contamination, but it is likely to be the consumer. As we wait to find out, it is worth looking at another part of the fall-out from the Mexican maize fiasco. Ignacio Chapela believes that one of the reasons he was attacked is because he had opposed the corporate of alliance between Berkeley and Novartis; that he had opposed the corporatization of science. But it is not only in the USA that it is happening.

CHAPTER 9

Science for Sale

*'The government have never put consumers or the environment first.
They have always put the industry and competitiveness first and that
is just clear in all the policy documents and statements'*

Dr Sue Mayer

*'If you go round the universities, there's a lot of entrepreneurial activity
going on. I don't think we have celebrated this important change'*

Lord Sainsbury, Science Minister[1]

The audience buzzed at Carlton House Terrace, the home of The Royal
Society. The scientific establishment had come to listen to Tony Blair, in
May 2002, give a major speech on why 'science matters'. Blair told the
audience that he had met a group of academics involved in biotechnology
in Bangalore in January. 'They said to me bluntly: Europe has gone soft
on science; we are going to leapfrog you and you will miss out. They
regarded the debate on GM here and elsewhere in Europe as utterly
astonishing. They saw us as completely overrun by protestors and
pressure groups who used emotion to drive out reason. And they didn't
think we had the political will to stand up for proper science.'

Blair delighted his audience by saying 'our science base is a key
national asset as we move into the 21st century'. One of the driving
forces was biotechnology which was 'at the forefront' of scientific
developments. 'The biotech industry's market in Europe alone is expected
to be worth US$100 billion by 2005. The number of people employed in
biotech and associated companies could be as high as three million, as we
catch up with the US industry – currently eight times the size of Europe's.
And Britain leads Europe.'[2]

The Blair government sees the biotech industry as the new 'scientific
frontier' offering 'potential massive benefits'. It will be central to the

future of the knowledge economy. The UK's lead in biotech cannot be compromised as the race to commercialize science continues unabated. Gone are the days of public sector research being undertaken for the public good. Over the last 30 years there has been a revolution in the way science is organized and funded.

As short-term profit becomes the engine driver of science, the long-term consequences are devastating. Fear stalks the corridors of academia, as people are afraid to speak out. Those who do, like Pusztai and Chapela, are hounded to the point that junior scientists know that speaking out is neither good for one's career nor for one's health. The direction of science is changed as profit becomes the overriding driver. Those who do not fit the bill get closed down or sidelined, whether they are doing pioneering research or not.

It could be argued that both of the agricultural disasters outlined in this book – BSE and foot and mouth disease – were a direct consequence of Britain's scientific policy over the last 30 years. It is hard to relate a farmer's suicide to scientific policy, but the links are there.

BSE scientists noticed the change in funding. When Professor Richard Lacey started his career in the late 1960s, he was a lecturer and reader at Bristol University, a post created by Harold Wilson. 'I was given the freedom to do what research I wanted to do'. By the time he moved to become a Professor of Clinical Microbiology at Leeds University in 1983, he noticed how 'completely different' the funding had become. 'Their only concern was getting money, and what had happened at the beginning of the eighties was the beginning of the withdrawal of central funding. Science now has changed from the search for knowledge in its own right, to the search for something that is commercially exploitable, which may not be much value to the human population. It is changing for the worse.'

In the early 1980s, a Policy Advisory Committee was imposed on the then Agricultural Research Council, who decreed a '40 per cent cut in research on animals diseases'. Would it be unfair, asks Alan Dickinson, one of the world's leading experts in scrapie (see Chapters 1 and 2), 'to mention the sequence since the early 80s? Swine fever, BSE, E. coli 157, foot and mouth'.[3]

Dickinson was the Director of the Neuropathogenesis Unit in Edinburgh from 1981 until he resigned in 1987. During that time, his research unit was increasingly squeezed of funds as the Tories began to attack the funding of science. TSE research – where experiments were lengthy due to the long incubation period – was diametrically opposed to the new quick fix, cash-it-in mentality that began to pervade science. If the NPU had been properly funded, had still had academic freedom and support, the BSE crisis might never have happened. Or if it had, it would have been totally different.

In the BSE epidemic, it was MAFF personnel who first identified meat and bone meal as the most likely source of the BSE infection. But Dr Dickinson, the first head of the NPU believes that this finding could have been reached a whole year earlier 'if the biosciences and TSE research in particular' was not engulfed in the turmoil of the Thatcher cuts and reorganisations. In short the BSE epidemic would have been 'considerably smaller'.[4]

So would have the foot and mouth disease epidemic. 'Really big errors in the 1970s and 1980s meant that two brilliant senior foot and mouth scientists were glad to leave the Pirbright institute' says Dickinson, 'the recent national crisis would have been different if they had not been lost'. One person to leave was Professor Fred Brown, who believes that it was the government's laboratory at Pirbright's commercial association that stopped the British authorities using a test that, he believes, could have prevented the contiguous cull. 'I think that a government lab should never be forced into a collaboration because of getting money, when it is a thing like this,' says Brown. 'A country that is as rich as Britain should not have to force its government laboratories to go cap in hand to a company in order to keep its thing going. This was the encouragement of the Thatcher years. Collaborate, collaborate, and take out patents. Absolute bloody rubbish. There is a point where you have to be independent, where you say: "No we are us, we serve the country, we don't serve the particular company".'

So the short-term and shortsighted mentality of successive governments to save cash, has cost over 100 people their lives through vCJD, millions of animals have been slaughtered and the country has lost some £20 billion.

But despite the consequences, since the 1980s successive UK governments have been trying to commercialize science and make it one of the principal forces behind the ubiquitous drive to improve the competitiveness of UK's industry. This is due to the long-held paranoia that 'ever since the country lost its industrial and technological leadership towards the end of the 19th century' the UK has been 'underperforming in creating wealth from its world class science base'.[5]

National wealth creation through the promotion of economic competitiveness is now the driving force behind science, with pressures on public sector scientists to produce commercial results.[6] The changes have been so radical that some argue that 'British higher education has undergone a more profound reorientation than any other system in the industrialised world'.[7]

By 1987, just as the BSE epidemic took hold, the Government's position had changed to 'favour support for strategic science and the promotion of university/industry links'.[8] Within five years the Tory government had set up the Office of Science and Technology to improve its management of science and technology policy.

By 1993, the most far-reaching review of science for a generation was published in the government's white paper 'Realising Our Potential: a Strategy for Science, Engineering and Technology'. It outlined the need to increase competitiveness by creating partnerships between industry, scientists and government; by identifying 'exploitable technologies', and focusing on 'market opportunities' in science. This strategy led to the introduction of the Technology Foresight Panels a year later and is embedded in the remits of the Research Councils.[9]

One of the primary goals of the Foresight Programme is to 'build bridges between business, science and government'. At the heart of the programme are a series of 13 panels, which are supported by a series of task forces that look at specific issues – one of which is the aptly named 'Food Chain and Crops for Industry' panel.[10]

But the Tory government had another radical idea too. In 1995, the Office of Science and Technology was moved from the Cabinet Office to the Department of Trade and Industry (DTI). 'What the change means is that we in Government recognize the vital need to bring industry and academia closer together', said the then science Minister.[11] So the Secretary of State for Trade and Industry has overall responsibility for the government's science policy, and is now supported by the Minister for Science – who is answerable to the DTI – and the Office of Science and Technology (OST), which is also part of the DTI.[12]

When Labour swept to power in 1997, it came with many ideas plucked from across the Atlantic. The British were envious of America's successful economy, and its booming technological industries as typified by Silicon Valley. The new Labour government wanted the UK's universities and research institutes to become one of the engine drivers for the UK's new economy. Technology was going to become one of the wealth creators of the 21st century and the innovation it would create would outshine our reliance on our dwindling manufacturing base.

The merger of business and science became a key part of New Labour's revolution for the private and public sector, and Labour carried on from the Tories. 'In the knowledge economy' says Tony Blair, 'entrepreneurial universities will be as important as entrepreneurial business'.[13]

Just over 18 months after coming into power, in December 1998, the UK government launched the White Paper 'Our Competitive Future: Building the Knowledge Driven Economy'. 'In government, in business, in our universities and throughout society we must do much more to foster a new entrepreneurial spirit: equipping ourselves for the long-term, prepared to seize opportunities, committed to constant innovation and enhanced performance' argued Blair.[14]

In the White Paper the government promised to vigorously promote the commercialization of university research – including new incentives for researchers to work with business.[15] Commercial exploitation was the buzzword in the report. The then DTI Secretary of State, Peter Mandelson, announced that the DTI's Innovation Budget would be increased to some £220 million to refocus the DTI's resources on partnerships between industry and science. 'We must encourage universities to engage with business so that new discoveries can be turned into marketable products as quickly as possible', said Mandelson.[16]

The same month the Higher Education Funding Councils (HEFC) for England, Scotland and Wales noted 'the spectacular growth in recent years across the United Kingdom in the scale, number and variety of linkages between higher education and industry'.[17]

The following year, 1999, just as the public backlash against GM foods reached fever pitch, it became clear how important biotechnology was to New Labour's goals.

The same month that The Royal Society and the government dismissed Pusztai's research, the incoming secretary of State for Trade and Industry, the Rt Hon Stephen Byers, MP, described biotech as 'an industry of the future – a knowledge driven industry – with great potential to create wealth and jobs. Exactly the sort of industry the Government wishes to promote'.[18]

Even The Royal Society, the 'bastion' of science is moving closer to industry. In the 1990s, 'supporting organisations' included Amerada Hess, British Gas, BP, Esso, Filtronic, GEC, Glaxo Wellcome, NatWest Bank, Nycomed Amersham, Rhône-Poulenc, Rolls Royce, UKAEA and Zeneca.[19]

Stephen Byers had made his comments to the House of Commons Environmental Audit Select Committee. The committee were told by the Cabinet Enforcer, Dr Jack Cunningham that the UK had 'formidable' interests in the GM sector and was a world leader in the industry, second only to the USA, with about 250 companies employing 14,000 people. The world market was forecast to reach £70 billion by 2000 of which the UK's share might be about £10 billion.[20]

In the summer of 1999, the Labour government released its 'Biotechnology Clusters' report, written by Science Minister Lord Sainsbury, who has been a controversial figure in the GM debate. Sainsbury has withstood numerous attempts by the Opposition to get him to resign because of his 'conflicts of interest' over his biotech activities and being involved in policy decisions that relate to biotech. Sainsbury is seen as one of the key people promoting biotechnology in

Labour's government. In 1987 he had set up the Gatsby Charitable Foundation, which donates about £2 million to the Sainsbury Laboratory[21] that is linked to the John Innes Centre. Both centres are leaders in biotechnology. Until they were placed in a blind trust when he became Science Minister, Sainsbury also held large share holdings in two biotech companies, Diatech and Innotech.[22] The irony of Sainsbury being in charge of a pro-biotech science policy at the same time as holding blind biotech stock was highlighted when Sainsbury made a £20m paper profit on GM food shares, in just four years, through his investment in Innotech.[23]

Lord Sainsbury's Cluster team concluded that the 'Government must do all it can to support the success story of the UK biotechnology industry and ensure that we maintain our lead in Europe'.[24] Sainsbury announced the establishment of the Biotechnology Exploitation Platform Challenge, the aim of which is 'to anchor the benefits of publicly funded bio science research in the UK'. It is worth some £6.4 million.[25]

Another major initiative was the new round of Foresight programmes designed to last until the spring of 2002. It was described as the 'most ambitious programme of future thinking ever undertaken in the UK'.[26] One of the Sectoral Panels was the aptly named Food Chain and Crops for Industry (FCCI) Foresight Panel. Six different task forces were set up,[27] including 'Unlocking the potential of GM Crops', whose members were heavily drawn from industry, including Zeneca and DuPont. On every committee, consumer representatives were in a small minority compared to industry and science.[28]

At the panel's third meeting in February 2000, they heard from the 'technology' sub-group on possible future food products, including 'Nutraceuticals', and GM crops tailored to individual health needs, such as the 'Viagra bun'. Other potential foods discussed included GM bananas in Scotland.[29]

That same month the influential Council on Science and Technology, which has the ear of the Prime Minster, published a report called 'Technology Matters – The exploitation of science and technology by UK Business'. It outlined how the 'strategically important sun-rise' industries must be nurtured and exploited – computing, microelectronics, telecommunications, advanced materials and biotechnology. It argued that the two-way flow between companies and universities needed to be increased. The message was simple: scientists and science graduates should 'maximize interaction with business'.[30]

In July 2000, the government responded to Technology Matters with the publication of its latest White Paper, concluding that that the UK 'could become a scientific hub of the world economy'. The White Paper talked about the 'new role' of universities and a 'change in culture', creating a climate for enterprise in all our universities. 'We must have the ability to generate, harness and exploit the creative power of modern science.' Science was no longer the discovery of knowledge, it was to be used as a means for companies to be able to 'create competitive advantage'.[31]

This exploitation of science has been continued ever since by Byer's successor at the DTI, Patricia Hewitt. In October 2001, Hewitt announced that Labour was giving £120 million for knowledge transfer funding: to transfer knowledge from the public to the private sector. The whole thrust was to create 'entrepreneurial education', to turn good science into good business and the commercialization of research. 'It is essential we have the links necessary to turn new ideas and technology into prosperity and jobs', said Hewitt.[32]

Two months later, Hewitt argued that 'our task now is to ensure that we get more of that science and technology out of the labs and into the factories. I want to see more "invented in Britain" become "made in Britain". That is how we will get the new industries and the new manufacturing jobs for the future and improve quality of life for everyone'.[33]

This theme was continued in the summer of 2002 with the publication of the government's latest strategy for science, engineering and technology, called Investing in Innovation. 'The 2002 Spending Review announced the largest sustained growth in science expenditure for a decade – £1.25 billion extra a year by 2005–06', noted Gordon Brown, Patricia Hewitt and Estelle Morris, in the introduction. 'For far too long British science has been denied the opportunity to develop. We now have the chance to turn this around: to make more British inventions become British manufactured products, creating jobs and prosperity for all.'[34]

One of the key proposals was to provide substantial new resources to the Research Councils from 2005–2006, to enable them to make a more realistic contribution to the full costs of the research that they sponsor in universities.

Some scientists argue that the Research Councils, 'as originally created, were well conceived to ensure that basic research should be funded so as to ensure its objectivity and freedom from coercion'.[35] But times have changed.

Realigning the Research Councils

The seven Research Councils that control much of the funding and direction of British science have moved from funding 'out of the blue' or 'blue sky' scientific ideas to ones that 'are consistent with the broad research priorities – areas of strategic science – identified by the government'.[36] So a centralized system of science exists, with industry and the government pulling the purse strings.

The leading funding agency for the biosciences at universities and institutes in the UK is the Biotechnology and Biological Sciences Research Council, which replaced the old Agricultural and Food Research Council in 1994, the year after the influential Realising Our Potential White Paper in 1993. As the BBSRC outlines, the White Paper 'made it clear that decisions on research and training support should be closely related to the country's needs and to enhancing the capacity to create wealth. Reflecting this emphasis, users in both government and industry make up half the membership of the BBSRC Council. All members are appointed by the Secretary of State for Trade and Industry'.[37]

The name change alone of this council signals the intent of the government, and the BBSRC, as it is known for short, now supports over 5000 researchers and research students at 60 universities throughout the UK and in eight BBSRC-sponsored research institutes. Some 35 per cent of BBSRC funds are allocated to its institutes, which mainly work on agri-food. The eight BBSRC-sponsored institutes are the Babraham Institute, the Institute for Animal Health, the Institute of Arable Crops Research, the Institute of Food Research, the Institute of Grassland and Environmental Research, the John Innes Centre, the Roslin Institute, and the Silsoe Research Institute.

The BBSRC is headed by Dr Peter Doyle who, until his retirement, was a senior executive with Zeneca. Its committees are packed with biotech enthusiasts, including people from Syngenta, the Roslin Institute, Genetix plc, GlaxoSmithKline, the John Innes Centre, Unilever, Biogemma and AstraZeneca.[38]

In February 2000, a group of senior scientists, including Ray Baker and representatives from Zeneca, Unilever, Glaxo Wellcome and the BioIndustry Association, developed 'a scenario for biotechnology' up to 2005.[39] Noting that public confidence with GMOs would hamper biotech applications in food, their outline indicated that there was still a potential for new markets such as 'nutraceutical' products,[40] the so-called second generation GM foods, whose safety worries so many people.

Nevertheless the scientists concluded that the importance of biotechnology would be greater after 2005, as the technology became 'more pervasive'.[41] It is not surprising that the BBSRC reached this conclusion as it actively encourages collaboration between scientists and industry, and its representatives come mainly from industry. In response to the government's White Paper on Science and Innovation in 2000, the BBSRC particularly welcomed its 'support for incentives for research staff to seek commercial development of their science, and for research bodies to own intellectual property rights'.[42]

Indeed the BBSRC aims to foster 'an entrepreneurial culture' in science through various awards and initiatives, for example 'Industrial CASE' studentships, where companies such as AstraZeneca, Aventis Crop Science, DuPont, Eli Lilly, GlaxoSmithKline, Novartis, Roche, GlaxoSmithKline, Unilever and Zeneca Agrochemicals have linked up with scientists.[43]

One dynamic area ready for expansion is spin-off companies set up to exploit research. Says the BBSRC: 'There have been many examples of spin-out companies arising from BBSRC-supported research including, in 1998 alone, MicroGenics (Oxford University), Biotica (Cambridge University) and Roslin BioMed (Roslin Institute)'.[44]

Many research institutes and universities have commercial arms. Take the Rowett, Pusztai's old employers. About 60 per cent of its income comes in grant-in-aid form from the Scottish Executive Rural Affairs Department. Other income comes from the EU and commercial sources, with 10 per cent coming from Rowett Research Services Limited, the commercial arm of the Rowett.[45] The Rowett's partner, the Scottish Crop Research Institute, has a commercial arm called Mylnefield Research Services Limited. These two are interlinked with the biotechnology company Biosource Technologies Inc in a deal seen by the SCRI as a 'major step in the advancement of plant biotechnology to revolutionise the treatment of mammalian diseases'.[46]

There are increasing links between leading research centres and biotech giants, which are welcomed by both government and the BBSRC.[47] In 1999, The John Innes Centre (JIC) signed a major deal with DuPont and Zeneca (later to become Syngenta), raising their industrial funding from 2.5 to 9 per cent.[48] As part of this collaboration, the John Innes Centre built a genomic science research centre, jointly funded by Syngenta, the East of England Development Agency (EEDA), and the BBSRC.

At the time, both Syngenta and the John Innes Centre heralded the joint venture as a major success. Syngenta called it an 'excellent example of a partnership between independent, publicly-funded science and a commercial organisation' that would 'benefit the UK/European science base as well as Syngenta, JIC and the Sainsbury Laboratory'.[49] This whole strategy backfired badly when Syngenta announced it was pulling out of the deal in September 2002. 'The frustration for us is this was a new kind of relationship that looked like setting an exciting precedent,' said Ray Mathias, from JIC.[50]

But just how far commercial pressures rule is shown by the way that some scientific results are now delivered to the City of London, rather than to peer-reviewed journals. PPL, one of the world's leading biopharmaceutical companies, announced in January 2002 that they had successfully cloned piglets, born on Christmas Day. However, the results were non-peer-reviewed and were announced first to the City, where PPL's share price rocketed by 44 per cent. The announcement came just two days before competing scientists in America published similar results in the American journal *Science*.[51]

'We're a public company and we decided to make a limited press release ... as soon as we felt that we had something [stock] price sensitive', Alan Colman, PPL's research director said. 'People don't have time to hang around and wait for a peer review.' Whilst PPL has been criticized by Philip Campbell, the editor of *Nature*,[52] The Royal Society and others that were so quick to punish Pusztai have remained oddly quiet.

What commercialization means for science

There is growing unease amongst many scientists, both in the UK and abroad, about the increasing commercialization of research and what this means for academia. 'I think there is a very real problem from the point of view of university research in the way that private companies have entered the university, both with direct companies in the universities and with contracts to university researchers,' says Stephen Rose, from the Open University. 'So that in fact the whole climate of what might be open and independent scientific research has disappeared; the old idea that universities were a place of independence has gone. Instead of which one's got secrecy, one's got patents, one's got contracts and one's got shareholders.'[53]

'Nowadays there is so much pressure on scientists', says Dr Milton Wainwright, a senior lecturer in the Department of Molecular Biology and Biotechnology at Sheffield University. 'The big difference now, compared to when I started in the business thirty years ago, is summed up by that word "money". When money comes into any system, when getting money becomes the sine qua non, then that's when the system gets corrupted. The emphasis on grant getting has corrupted science. Big science looks after itself. It is the mavericks that lose out.'

Speaking out in science has now become a dangerous thing to do. 'When people speak out, often they just don't realise how dangerous it is', says Dr Brian Martin, who is an associate professor in Science, Technology and Society at the University of Wollongong in Australia and author of *The Whistleblower's Handbook*. He has studied the suppression of scientific dissent in pesticides, nuclear power, fluoridation, forestry and works with whistleblowers on a daily basis. 'Many workers who become whistleblowers are hard-working, conscientious, and believe in the system', says Martin. 'They report some problem in good faith and suddenly they are harassed, abused, threatened, dismissed and they don't even know what hit them.'

Martin adds that 'The most insidious thing is the inhibition of dissent due to the lure of research grants and promotions and the inculcation of the commercial ethos within universities – that process has certainly increased. That is where the major change will have been occurring.'

Two headlines from the time of the Pusztai affair are revealing of why Pusztai received the treatment he did. The first, from *The Daily Telegraph*, says 'GM food scares "risk Britain's lead in science"' and reads that 'Britain risked losing its lead in one of the important industries of the 21st century'.[54] The second, from the *Aberdeen Press and Journal*, was entitled 'Pusztai accused of costing country millions over GM', quoted Professor Wilson, whose institute, the Scottish Crop Research Institute was one of the original collaborators in the GM potato project. Wilson maintained that Pusztai's comments had cost the world-wide food industry millions of pounds worth of damage.[55]

Dr Horton, the Editor of *The Lancet*, in a so-far unpublished interview for *The Guardian* in 1999, agreed that the whole saga was to do with money. 'In my view it is all about money for the food biotechnology industry' he says. 'If you look at the people who really said something aggressive about us publishing this, it's either executives from The Royal Society or plant biologists because they are the ones who are going to lose out.' Horton described scientists as going 'psychotic' because they see a massive source of funds going down the 'tube'.[56]

Horton's views are shared by others in the scientific community: one Royal Society Fellow, who wishes to remain anonymous, argues that biotechnology is one of the few industries 'we have a chance of doing well in'. There are 'people within the RS who are connected with industry who feel that if someone is doing anything that endangers that, then they are doing something wrong. My own feeling is that it is as much a commercial thing as a scientific thing as the scientific arguments don't add up'.

As we know, Chapela was a major opponent against the proposed $50 million deal between Novartis (later merged to form Syngenta) and the University of California at Berkeley, and he believes that his treatment with the Mexican maize fiasco reflects this. At the end of 2002, it looked like the controversial association between Syngenta and Berkeley might be coming to an end, after the renewal deadline passed with no new offer from Syngenta.

Chapela's unease was reflected by the Royal Society of Canada's panel of experts into biotechnology. They concluded that 'researchers have noted that these commercial alliances can have a profound impact on the choice of research topics. They also help to create an atmosphere of secrecy among researchers and jeopardize the trust which the public places in academic science'.[57]

They noted that 'the increasing domination of university research by the commercial interests of the researchers and their industry partners removes incentives for reliable scientific research on the safety of these products [GM]'.[58] The scientists concluded that 'the co-opting of biotechnology science by commercial interest contributes to the general erosion of public confidence in the objectivity and independence of the science.'[59]

In the UK, Professor Brian Wynne from the Centre for the Study of Environmental Change at Lancaster University, has conducted government funded research on social attitudes to GM. 'There is the sense that the whole of science is hijacked by private interests' says Wynne. 'Intellectual property rights and patenting are being pushed heavily and there isn't such a thing now as public interest science and that is part of peoples' deep mistrust of the regulatory system… People are very sceptical and they are animated by the perception that over a 15–20 year horizon all the pressure on research has been pushing it into the hands of industry.'[60]

Corruption of integrity?

There is evidence that corporate funding is contaminating research. In November 1999, the Institution of Professionals, Managers and Specialists (IPMS), a union that employs some 75,000 scientists and technicians in the private and public sector, published a survey called 'What Future R&D'. The survey revealed that 30 per cent of scientists had been found to tailor their research conclusions or resulting advice with 17 per cent asked to tailor their work to suit the customer's preferred outcome, and 10 per cent asked to do so to obtain further contracts.

The survey also found that 34 per cent of scientists thought that commercial spin-off companies were a 'negative development' on public sector science. It concluded that: 'an approach based solely on the commercial value of research activities cannot hope to do justice to the 'public good' value of science or to reward fairly the scientists that work in this area'.[61]

In March 2001, IPMS warned that their members were worried that 'objective and independent R&D is being placed under threat by increasing commercial pressure and that market forces are taking precedence over professional standards'.

The IPMS, which has subsequently merged with the Engineers' and Managers' Association to form Prospect, highlights the case of scientists who have suffered. One case involved an experienced scientist, called 'Joe', who was moved from studying climate change into research on genetically modified crops. According to the Union, 'Joe recognises the potential uses of GM plants in enhancing the level and quality of production and in resistance against pests, diseases and adverse environments. However he also believes that there are many technical, safety and ethical issues that still need to be addressed.' In a case startling in its resemblance to Pusztai, when Joe wrote a paper balancing the positive and negative aspects of GM 'his research group was disbanded and his management responsibilities withdrawn'.[62]

The IPMS has hosted a number of conferences that have examined science, scientific advice and its increasing commercial future. Another conference called 'Corruption of Scientific Integrity? – The Commercialisation of Academic Science', was held at the British Academy in May 2001. 'Down what river is academic science being sold?' asked Professor John Ziman, a Royal Fellow and Emeritus Professor of Physics at the University of Bristol. Ziman argued that at the beginning of the 20th century that there were two different forms of activity – science and industry – with governments and research councils in between. Since the 1950s, he contended the two have increasingly merged to create 'post-academic' science, where universities are undertaking commercial science. Ziman believes that this trend makes the scientist into a serf or even a slave, a mere instrument to serve the purposes of commerce.[63]

The next speaker was Professor Nancy Olivieri, who is a professor of Medicine at the University of Toronto and Hospital for Sick Children, an award-winning specialist in the treatment of hereditary blood disorders, especially thalassemia, a haemoglobin disorder. After alerting patients to a problem with drug trials, she endured six years of hell, repeatedly losing her job and being sued for $10 million by the drug company involved, Apotex. An investigation into the whole affair by the Canadian Association of University Teachers (CAUT) is a chilling read for anyone worried about the state of science and medicine.[64] 'Commercialization of university research', said Olivieri 'benefits companies at the expense of the public good.'[65]

Nancy Olivieri speaks with a passion burdened by her ordeal, which shows the fundamental failings of the corporatization of science. 'Emotionally, I'm broke,' she says. 'First I was fired, then I was not fired, because they didn't mean fired, then they said now you are fired again, then came a letter saying take medical leave from the hospital, and then there would be another attack from Apotex... I did lose my job four times.'

The CAUT concluded that part of the problem of the Olivieri saga had been due to the 'increased pressures on universities, teaching hospitals and individual researchers to seek corporate sponsorship for projects'. They wrote that 'unless the lessons are learnt everyone will lose. It is important to recognize that the circumstances that gave rise to this case are not isolated—they illustrate a system-wide problem'.[66]

Olivieri sees a parallel between what happened to her and Dr Pusztai. 'There is obviously no ethics that drives these people apart from money. They are trying to crush Pusztai. But it's built on companies and it's built on money. It's not built on what is right. Pusztai is up against a very big scientific machine. Which is really why I think his story is so lamentable, because The Royal Society didn't even have to put their hands above their heads, they just had to say "it was flawed". It is such a sell-out.

Olivieri is not the only academic to suffer at the hands of the University of Toronto. David Healy, author of the *Anti-Depressant Era*, had his job offer at the university turned down after he linked the taking of Prozac to suicide. Although the Prozac manufacturer, Eli Lilly, finances the university's Centre for Addiction and Mental Health, the university said that it was not influenced by outside forces.[67]

Healy has also been the victim of another by-product of commercialization: ghost-writing by companies. In September 1999 he was approached by a public relations company working for Pierre Fabre. He was invited to speak at a company-sponsored conference and his speech was to be submitted later for publication in a scientific journal. He was surprised when he received an email from the company saying that 'in order to reduce your workload to a minimum we have had our ghost-writers produce a first draft'. Healy refused this draft and wrote his own, but the agency submitted the article in another scientist's name as they needed 'to bring across' the main commercially important points' that 'are not accentuated in your manuscript.'[68]

A survey by *The Guardian* newspaper had found that ghost-writing had become 'widespread' in certain areas of medicine. Ghost-writing for companies and receiving fees for speaking was so common that one scientist comments that 'some of us believe that the present system is approaching a high-class form of professional prostitution'.[69]

There are academics who have suffered for speaking out. Dr Betty Dong, from the University of California San Francisco was involved in a study of Synthroid, the third most dispensed drug in the USA, taken by about 8 million people a day. Prescribed for thyroid conditions, it was produced by Boots, before the company sold its drug division to a subsidiary of BASF, called Knoll Pharmaceutical. When Dong found that the drug was not as effective as the company claimed, Boots launched a campaign to stop her from publishing her results.[70]

David Kern was a professor of Occupational Medicine at Brown University. One of his research projects was to look into the illness of two workers at a textile plant. His concluding diagnosis was that they were suffering from a new disease, called 'Flock workers lung'. When he attempted to publish this information, he was threatened with a non-disclosure agreement he had signed. He continued to go ahead and published in the *Annals of Internal Medicine*. Subsequently, he was fired by the university and his unit disbanded.[71]

In April 2001, *The Lancet* highlighted how the pharmaceutical company Bayer wanted to exert some degree of control over publication of research that concerned one of its drugs. The journal went on to note that '*The Lancet* recently came under pressure to remove a sentence from the discussion of a research paper, which raised questions over the safety of a drug. As research becomes driven by ever more costly technologies, so industry will intrude even further into the scientific process'.[72]

One of the scientists to support Ignacio Chapela over the Mexican maize saga, was Dr Allison Snow from Ohio State University. Although respected on both sides of the debate, Dr Snow fell foul of the biotech industry too. Snow's research was originally funded by Pioneer Seeds and Dow, together with the US Department of Agriculture. When Snow produced preliminary evidence that wild sunflowers, containing the transgenic *Bt* gene increased seed production, which could allow the plants to proliferate as superweeds, the companies were accused of blocking further research by denying Snow access to the transgene and sunflower seeds. 'It is very frustrating,' said Snow, who was said to be shocked by her results. 'We want to do good science. But this is keeping us from answering questions we want to ask.'[73]

It also happens in the UK. A fierce battle surrounds the safety of the triple MMR (measles, mumps and rubella) vaccine and inflammatory bowel disease/autism. In 1998 Dr Andrew Wakefield suggested a link between MMR and autism and caused a huge controversy, leading him to be attacked by Britain's medical establishment. He finally resigned from his post in 2001 under pressure from the Royal Free and University College Medical School. 'The hierarchy of the medical school decided it did not wish the work to continue there', Wakefield told *The Lancet*. 'I can only assume it [the research] was politically incorrect.'[74]

By January 2002, just as the MMR debate continued to rage in the British press, the Policy Commission on the Future of Farming and Food issued its findings. It recognized that there was a 'tension between the food and farming industry's need to adopt new technologies in order to remain competitive, and the nervousness of citizens who feel that their safety, or that of animals or the environment is taking a back seat to those economic goals'. The resolution of this tension will not be easy, felt the Commission, but it was 'essential for the harmonious future of the food industry and society'.

Another recommendation by Curry's team was that the government might want 'to make sure that the outcome of all private research on GM is, and is seen to be, subject to the most rigorous peer review'.[75] Even peer review, however, is not infallible. Marcia Angell, former editor of the *New England Journal of Medicine*, wrote in 2001 that when she ran a paper on an antidepressant drug treatment, the authors' financial ties to the manufacturers – which the journal requires all contributors to declare – were so extensive that she had to run them on the website.

Angell decided to commission an editorial about this event and spoke to research psychiatrists, but 'we found very few who did not have financial ties to drug companies that make antidepressants'. She wrote: 'Researchers serve as consultants to companies whose products they are studying, join advisory boards and speakers' bureaus, enter into patent and royalty arrangements, agree to be the listed authors of articles ghost-written by interested companies, promote drugs and devices at company-sponsored symposiums, and allow themselves to be plied with expensive gifts and trips to luxurious settings. Many also have equity interest in the companies'.[76]

The degradation of science

So what does all this mean? 'The bottom line for universities that they haven't fully understood', says Drummond Rennie, the deputy editor of the scientific *Journal of the American Medical Association* 'is that in the end, public universities have to rely on public support. If the public perceives a university as a place where scientists become millionaires and where companies are in control, they'll lose public support, and that will be catastrophic for them and for the public at large'.[77]

'The piper is calling the tune and it raises worrying issues' adds Charles Harvey, from the Institute of Professionals, Managers and Specialists. 'We have seen the BSE crisis, food scares and the GMO debacle and the public is losing confidence in government as an independent, fair-minded arbiter.'[78]

But not only has trust evaporated, but many argue that the commercialization of science is damaging the very process of science itself. 'What is amazing is that science still has as much credibility as it does given that commercial factors and issues of national security are so embedded in large slabs of science', says Dr Brian Martin from the University of Wollongong in Australia. 'At the moment a small number of people who have money and power are calling the shots on what research is done, and they don't like it if it serves other groups. If there was a democratic process of setting scientific priorities, research would be less slanted to particular interests.'

Steven Rosenberg from the US National Cancer Institute also takes issue with what is happening: 'As biotech and pharmaceutical companies have become more involved in funding research, there's been a shift toward confidentiality that is severely inhibiting the interchange of information', he says. 'The ethics of business and the ethics of science do not mix well. This is the real dark side of science.'[79]

The dark side of corporate control is that the scientific process itself is degraded. 'We are corporatizing science, turning scientists into accountants; we are trying to control it too much', argues Dr Milton Wainwright from Sheffield University. 'We often forget that one of the major forces behind paradigm shifts is serendipity, chance discovery.'

'No one's telling you that cars are one hundred per cent safe, but government and the public want scientists to do that on issues like MMR. When scientists pretend they know all the answers, then that's when things go wrong. That's when the public get really annoyed because they've heard it all before when things did go wrong. They mutter things like "of course they reassured about BSE, didn't they?" "Why do we pay scientists if we don't want them to tell the truth?"'

CHAPTER 10

Whitehall Whitewash

'We always put consumers' interests first'

> Suzi Leather, former Deputy Chair of the
> Food Standards Agency[1]

*'We remain bitterly disappointed at the anti-consumer stance the Food
Standards Agency and UK government, as a whole, take on this
issue* [GM labelling and traceability]*'*

> Sue Davies, Consumers' Association[2]

The Department for Environment, Food and Rural Affairs

For a ministry that many blamed for the BSE crisis, foot and mouth was the final straw for MAFF. The dreaded ministry was replaced by Defra, the Department for Environment, Food and Rural Affairs. Agriculture had been consigned to the dustbin whilst environment, food and rural affairs had risen to the fore. There were rumours that 'Food' had only been added at the last minute to the department's name, an oversight that cost £20,000.[3] The government said it would 'spearhead a major new drive on green issues and the countryside'.[4]

On 8 June 2001, the new department was born. 'It will adopt a truly joined-up approach to all aspects of our environment to ensure a high quality of life, vibrant and sustainable rural communities and a food chain that works together to meet the changing demands of consumers', said the new Minister, Margaret Beckett.[5]

Despite the rhetoric, the response was sceptical. 'Initial soundings from government suggest its definition of sustainable may be to "don't go bankrupt" – or larger, more "competitive" farms', wrote *The Guardian*.[6] 'We hoped we would get a good steak out of the reforms,' said one MAFF official. 'But we ended up getting the whole cow'.[7]

Seasoned MAFF watchers see little difference with the arrival of the new department. The website called 'cullmaff.org' argued that 'MAFF may have been culled but its vile policies live on in Defra'.[8] 'Can a leopard change its spots?' asks Arthur Beyless, who lost his daughter to vCJD. 'Even after the narrow escape the officials at MAFF had in getting away with being part of the causing of over 100 deaths of innocent people, including my daughter Pamela, they still come out with the same SPIN on the truth as they did with BSE'. The civil servants at Defra are basically the same as MAFF, says Beyless.[9]

The head of the civil service in the new department was Brian Bender, the ex-Permanent Secretary at MAFF and a favoured New Labour official.[10] Bender explained to the Select Committee on Environment, Food and Rural Affairs that basically his department moved with him. 'Numerically we are talking about 650 people from the former DETR, a handful from the Home Office and the entire staff of what was MAFF merging into a single department.'[11] Bender also conceded that ex-MAFF personnel constituted the biggest block on the new board.[12]

The day after the 'Lessons Learned' Inquiry into foot and mouth disease was published, *The Daily Telegraph* noted that Bender was one of the key characters involved in handling the disease. It read 'Brian Bender: Maff/Defra Permanent Secretary: Failed to see that ministry had a contingency plan in place for more than 10 cases... Presided over "a culture predisposed to decision taking by committee with an associated fear of personal risk taking"... His department had a "silo mentality". Defra's "patchwork of unconnected systems" hampered intelligence on the spread of disease. His department's "financial control systems were not up to the task"'.[13]

For some it was incredible that Bender, whose department made so many errors during the foot and mouth disease crisis, was put in charge of the new department. Christopher Booker and Dr Richard North wrote in *Private Eye*'s special edition on foot and mouth: 'On closer examination, MAFF was still there, very much as before; the same officials, in the same buildings, administrating the same policies'. In fact the MAFF to Defra makeover has been described as the 'most cynical makeover since Windscale changed its name to Sellafield'.[14]

So the culture remains and so do the denials that anything is wrong. When he appeared in front of the Select Committee on Environment, Food and Rural Affairs in November 2001, Bender was told that 'there did look to be a cultural problem within MAFF which also emerged from

BSE in the sense there is an introspective culture and culture of secrecy in that it does not like accepting outside help. Has this been tackled?' Bender replied that: 'I recognise the image you are talking about but I do not recognise it as a reality'.[15]

But probably more critical in the long term rather than the continuity of the personnel, was the continuation of the Whitehall structure and mentality. Sir Peter Kemp was the second Permanent Secretary in the Cabinet Office until 1992. He argues that instead of radical reform of MAFF 'all we got was a relatively minor reshuffle bringing the culture of agriculture and environment into conflict', but 'because the old MAFF culture was the stronger' very little has changed. Kemp argues that the MAFF/Defra model is exactly what not to do in reforming government. He believes there needs to be a new structure of departments, with a culture appropriate to the 21st century, rather than to those that date to a post-war mentality.[16] The Institute of Public Policy Research think tank articulated that reform was not 'simply about reengineering state bureaucracies'.[17]

Environmentalists and conservationists believe little has changed. 'The word already from within MAFF is that the ministry is not being broken up', said Ian Willmore from Friends of the Earth.[18] 'Defra looks suspiciously like MAFF with a little green knob tacked on,' added *ECOS*, the magazine of the British Association of Nature Conservationists.[19] 'MAFF may have changed its name to Defra, but civil servants remain the same – they have the same dogmatic views', adds Dr Elaine King from the National Federation of Badger Groups.

'Defra is not fully fit for its new purpose,' said Fiona Reynolds, head of the National Trust. 'It has been quick to develop direction, but it has been slower in delivery. Delivery on the rural white paper and the future of farming requires a shift in culture. It has a vast array of public bodies and agencies to inform, guide and deliver its policies, but many of these reflect the needs of the past – such as boosting agricultural production – and not of the future.'[20]

The minister in charge of the new department was Margaret Beckett, who is seen as hostile by farmers' leaders.[21] One of her first tests was whether to decide to order a public inquiry into foot and mouth disease. She refused, despite calls from the NFU, the Soil Association and bodies such as the Royal College of Veterinary Surgeons. Then came Defra's Animal Health Bill that has been nicknamed the Animal Death Bill. It is seen as a draconian piece of legislation that penalizes both animals and farmers. Soon Defra's critics were calling it the Department for the Elimination of Farming and Rural Affairs.

Six months after its inception, Defra produced a booklet called 'Developing Defra – Six Months on'. Government critics were not impressed. 'There is nothing in the document about what the real aims of Defra are, only management newspeak', said Conservative MP Keith Simpson for Mid-Norfolk. 'The document is meaningless… I fear that the new structure is not based on lessons learned from the past. It will not deliver what people want; instead, it will disillusion them with clichés and soundbites.'[22]

Just under a year after its inception, Defra tried to silence its critics again, producing its departmental report in May 2002. 'By bringing together the areas within Defra we have improved the structures that are essential to the government's programme on environment, food and rural affairs', argued Margaret Beckett.[23] But the Environment, Food and Rural Affairs Select Committee dismissed the report, saying it contained too much 'waffle', and 'next to useless' data. 'The presentation of the Defra departmental report leaves a lot to be desired', concluded the MPs.[24] 'We are thoroughly dissatisfied with the department's annual report. The problem is that there are serious omissions and it is riddled with inaccuracies. It also contains too many warm words and vague aspirations, and too few real figures against which its performance can be measured', said the chair David Curry.[25]

By October 2002, Defra's 'Sustainable Agricultural Strategy' was nearing completion. Once again seasoned observers were wary. 'Sustainable agriculture is such a new concept within a government department that for decades has outshone the world in unsustainable agriculture,' wrote Jonathan Porritt, Chair of the UK Sustainable Development Commission, 'as to make one just a little nervous as to what is about to appear'.[26]

And here lies the problem. Defra's predecessor, MAFF, was a department that was totally unsustainable. It pursued a policy of increased productivity at all costs, a course that led to the destruction of farming jobs and despoliation of our landscapes and wildlife. It all led to 'an edifice of bureaucracy and subsidy, which will be nearly impossible to dismantle'.[27] To all intents and purposes that edifice is still there, untouched, no matter what its new name.

The Food Standards Agency

The main positive legacy of the BSE crisis is undoubtedly the formation of the Food Standards Agency, headed by Sir John Krebs, which was set up in April 2000. The FSA has to be seen as a major step forward in relation to safe food. But how big a step is it really?

Just as Defra produced a one-year report in 2002; the FSA produced a two-year update on its activities. 'Our independence is vital if we are to succeed in putting consumers first', read the introduction to the report. 'Uniquely among Government departments, we can publish our public health advice without the agreement of Ministers.'[28]

Ministers like to call the FSA independent. It is right, says Margaret Beckett, that 'the Food Standards Agency, which is very much an independent agency and an independent voice in government, should report to the Department of Health'.[29] But just how independent is the FSA? Professor Philip James drew up the blueprint for the FSA, and was tipped to land the coveted top job, until the Pusztai controversy damaged his chances. 'The ministers were completely behind the blueprint but the civil service was desperately trying to claw back control into the civil service,' he says.

Professor James argues that there were two key decisions that went against what the blueprint had proposed. First, the independence of the FSA. 'When you look at the way the FSA was organised they managed not to make the staff independent of the civil service which we'd identified as critical for establishing its independence', says the Professor. Secondly, argues James, 'they appointed senior MAFF staff to the senior echelons of the agency when I'd made it quite clear from our analysis of previous experience with Health and Safety that you needed to bring in outsiders'.

James argues that when staff involved in drafting the legislation 'suddenly saw the final decisions' being 'controlled by MAFF' they 'immediately asked for a transfer because the FSA became a semi-governmental department'. So, argues James, 'if you look at the way in which it's now working it's far more constrained than I intended'. He also contends that anyone who had been prominent in the food debate and 'knew anything about the problems' was 'automatically removed from the shortlist' to the council and the board. James gives the Agency marks of six out of ten on their performance so far.

With James out of the running, Sir John Krebs became the FSA's first head. Sir John joins the list of members of the Zoology Department at Oxford, whose speciality is not primarily farming nor food, who have taken key positions on government committees or within government that seem to be outside their area of expertise.

Krebs was most well-known for his design of the 'Krebs experiments' to investigate whether badgers are responsible for increasing incidences of TB in cattle. These experiments have lead to the slaughter of some 20,000 badgers, according to the National Federation of Badger Groups (NFBG).[30] The NFBG believe that the way that Krebs worked with MAFF on what many believed to be a 'flawed' experiment, showed that he was willing to toe the line with MAFF, not stand up to it.[31]

'Vets within MAFF had for years wanted to carry out a massive badger culling experiment and Krebs simply rubber stamped it. It wasn't necessarily his idea and he didn't question the fact that his remit forced him to focus solely on badgers,' says Dr Elaine King from the NFBG. 'The report did not, for example, address cattle husbandry or cattle to cattle transfer of TB, something that was criticized by the Agriculture Select Committee when it investigated the issue in 1999.'

If Krebs had been controversial before his appointment, then things have got worse since. On the day it was announced that Krebs was becoming the head at the FSA, he was endorsing GM food, saying GM products 'were as safe as their non-GM counterparts'.[32] He was repeating his public position that criticisms of GM food were 'shrill, often ill-informed and dogma-driven'.[33] Indeed, Harry Hadaway, then at the Soil Association, called Krebs 'an historic supporter of GM foods', a position that the Government knew he had before offering him the post at the FSA.[34]

Shortly after joining the FSA, Krebs also aligned himself with an Oxford-based organization known as SIRC, the Social Issues Research Centre, which has set itself up as an arbiter of what is good and bad in the journalistic reporting of health and science stories. Sir John Krebs, SIRC, the Royal Institution, The Royal Society and others developed a set of 'Guidelines on Science and Health Communication', a revised version of which was published in November 2001.[35]

The involvement of SIRC is peculiar. SIRC maintains a pro-biotech position, 'taking into account the potential benefits of GM technology in disadvantaged areas of the world'. SIRC's funding comes not only from its 'sister organisation, MCM Research', but also from the Ministry of Defence, several large food companies, and the drink industry front organization, the Portman Group. It shares offices, directors and key personnel with its sister organization, MCM Research, a PR company whose clients include the Who's Who of the international drinks industry, and Conoco, the oil company.[36] 'How seriously', asked the *British Medical Journal*, 'should journalists take an attack from an organization that is so closely linked to the drinks industry?'[37]

Sir John Krebs is also on the Science Advisory Panel of the Science Media Centre (SMC), an 'independent organisation', whose funders include BP Conoco, DuPont, Tesco, and Astra Zeneca.[38] Instrumental in its formation was Susan Greenfield, director of the Royal Institution, which also helped launch the £120,000 centre.[39] The Centre was accused of 'orchestrating a secret campaign aimed at discrediting' a programme called *Fields of Gold*, a fictional thriller highlighting the dangers of GM, co-written by Alan Rusbridger the editor of *The Guardian*.[40] The majority of scientists 'offered up' by the SMC to speak to the media were pro-GM. 'It's come to something', wrote anti-GM activist Jonathan Matthews, 'when such a partisan group and such propagandist opinions are presented to the media as the "voices and views of the scientific community"'.[41]

But Krebs' most controversial stance so far has been on organic agriculture. Appearing on BBC TV in August 2000, Krebs announced that consumers who were buying organic food were 'not getting value for money, in my opinion and in the opinion of the FSA, if they think they are buying extra nutritional quality or extra nutritional safety, because we don't have the evidence.'[42]

A month later Dr Patrick Wall, the chief executive of the Irish counterpart agency, the Food Safety Authority of Ireland, dismissed Kreb's views as extreme and reminded people to buy organic food because it was more 'environmentally friendly, more wholesome, and better produced'.[43]

In March 2002, Krebs was criticized by John Paterson, a biochemist at Dumfries and Galloway Royal Infirmary, for attacking organic agriculture 'on the basis of very little information'. Paterson and a team from the University of Strathclyde found that organic vegetable soups contain almost six times as much salicylic acid as their non-organic counterparts. The acid is responsible for the anti-inflammatory action of aspirin and helps combat hardening of the arteries and bowel cancer.[44]

That autumn it was revealed that Krebs had been refusing to back the government's drive to promote organic food and farming, prompting the Environment Secretary to write to him to clarify his views. Sir John also admitted that comments he made that manure caused more air and water pollution than chemical fertilisers had been designed to undermine claims that organic farming is more environmentally friendly than conventional agriculture.[45]

Whilst other far more important issues such as BSE lay unresolved, many people in the food debate were concerned that Krebs and the FSA had effectively attacked organic agriculture. Professor of Food Policy from Thames Valley University, Tim Lang, says he feels that Krebs was 'being set up to say things he doesn't believe and the evidence doesn't warrant'.[46]

The real question is have Krebs and the FSA fallen victim to a long-standing campaign by agribusiness and right-wing think tanks in both the USA and the UK to undermine organic agriculture? Many of the same organizations that promote biotech also denigrate organic and you only have to look at the website of the AgBioWorld.org Foundation to see the latest vitriolic attack on organic agriculture.

To understand the answer, it's necessary to understand not only what organic agriculture stands for, but also what it stands against. The organic movement, based on a pesticide-free philosophy, seeks a more sustainable and holistic agricultural system; one that is fundamentally opposed to biotechnology. Organic agriculture is the biggest obstacle in the way of the biotech revolution.

'Agribusiness companies were perfectly happy to ignore organics when it was a tiny niche market,' says Jeanette Longfield from Sustain, the UK alliance for better food and farming. 'Now it is no longer a niche market they are clearly thinking it is going to have an impact on profits and they had better do something.'

That 'something' is an increasingly ruthless attempt to destroy the organic movement. 'The agribusiness companies are taking a two-pronged attack' says John Stauber, from *PR Watch*, an investigative quarterly in the USA. 'Firstly, big businesses are buying up organic processors and marketers. Secondly, at the same time, these companies are blasting the integrity of organics through their PR front groups. It's a brilliant 'win win' strategy for business.'

In the UK, the strategy is similar. Either they attack it: 'I think the FSA were completely right in what they were saying in that organic food is no more nutritious than conventionally grown food', say Novartis. Or they want to co-opt it: 'The aims and objectives of the people who are producing organic foods are very similar to ours,' says Professor Howard Slater, a spokesperson for CropGen. 'We would be very keen to see organic farming take on some of the GM crops that are beginning to become available and to use them within their regime.'

Both strategies will lead – quite deliberately – to the undermining of the organic symbol. 'One of the dangers for the organic movement is the appropriation of its ideology by big business and I think that is happening', says Dr Ben Mepham, from the Food Ethics Council. 'The word organic may not mean that much soon.'

In these efforts, the agribusiness and biotech corporations in the USA and the UK are supported by a loose network of right-wing think tanks. To an unsuspecting eye, these think tanks appear to offer a veneer of independence from the big businesses that financially support them to push forward a deregulatory, pro high-tech, corporate agenda.[47]

The main person behind many of the attacks on organic food is Dennis Avery, the author of the inspirationally-titled *Saving the Planet with Pesticides and Plastic: The Environmental Triumph of High-Yield Farming*, Avery sees himself as a missionary, promoting the high-tech farming industries: pesticides, irradiation, factory farming, and the newcomer: biotechnology.[48] He 'welcomed' Krebs' 'well-considered' attack on organic foods.

Avery, a former agricultural analyst for the US State Department during the Reagan era, is now Director of the Center for Global Food Issues, part of the Hudson Institute, a right wing US think tank. Avery's message is simple: organic food takes up too much land, and is actually dangerous for you. The growth in organic agriculture is due to an 'image created by the environmental movement'. It is a 'gigantic marketing lie'.

Avery believes that 'Genetically modified foods are significantly safer than organic and natural foods. Over the last decade, consumers have eaten millions of pounds of genetically altered foods, and millions of tons of feed corn and soybean meal have been used to produce our meat and milk. So far, not even a skin rash has been linked to these new-tech foods'.[49]

Harry Hadaway, formerly of the Soil Association, responded that Hudson's comments are 'scientifically unsound,' arguing 'the protagonists of GM and those involved in the Hudson Institute are keen to promote the use of any technology which will improve the financial position of the companies backing them'.

Avery dismisses critics who point out the funding of the Hudson Institute by agrochemical companies, by saying that 'I am not bought. I am a missionary'. But Hudson's Board includes current or ex-employees of Burson-Marsteller and Hill & Knowlton[50] both of whom have a history of working to counter environmental goals. Hudson's funders include many companies behind the agribusiness and biotech revolution: Ciba-Geigy (now Novartis), Cargill, Dow Elanco, DuPont and Monsanto.[51]

As the attacks on organic food increased, so others at the Hudson Institute joined Avery in his anti-organic fight. Other officials at its Centre for Global Food Studies include Avery's son, involved in the anti-Chapela campaign on AgBioView.org.

There is a cross-pollination of people, ideas and articles between the USA and like-minded right-wing think tanks and academic institutions in Europe. These contrarian groups and individuals deliberately reiterate each others' work in order to generate a critical mass of contrarian thought, which is picked up by a media anxious to find opposing viewpoints on previously uncontentious issues.

The strategy has worked before; namely with climate change, when views from a small group of scientists funded by the fossil-fuel lobby were repeated so frequently that they were given far more prominence than their unsupportable theories actually deserved.[52] In attacking the organic movement, the contrarians are using the same tactics, backing their arguments up by quoting the same small group of right-wing groups or corporate funded scientists.

One of the central characters spreading the anti-organic backlash in Europe has been Roger Bate from the Institute of Economic Affairs (IEA), which is one of the UK's leading right-wing think tanks. Bate co-founded the European Science and Environment Forum (ESEF) in 1994. ESEF was formed, in its own words, as an 'independent non-profit making alliance of scientists whose aim is to ensure that the environmental debate is properly aired… To maintain its independence and impartiality ESEF does not accept outside funding from whatever source'.[53]

In *Trust Us, We're Experts,* John Stauber and Sheldon Rampton outline how by 1994 Philip Morris had budgeted US$880,000 to fund a front organization on science in the USA. In consultation with Burson-Marsteller, the company also planned a second European organization, tentatively named Scientists for Sound Public Policy. Burson-Marsteller's documents showed that 'a countervailing voice must be created in Europe' with support from tobacco, agri-chemical, pharmaceutical, and biotech companies amongst others. Scientists for Sound Public Policy never materialized, although 'it appears that the outcome was the European Science and Environment Forum (ESEF), established in 1996, whose executive director sought funding from the tobacco companies', argues anti-tobacco academic Dr Stan Glantz.[54]

In August 1999, a book called *Fearing Food; Risk, Health and the Environment* was published, edited by Bate and a colleague from the IEA, Julian Morris. 'The book shows that intensive agriculture is good for health and the environment, and is essential if the world's population is to be fed without converting vast areas of biodiverse ecosystems into cropland, which would be necessary if organic agriculture, with its lower yields, were used', said the press release.[55]

Roger Bate and Julian Morris have recently also expanded their activities as 'staff' and Co-Directors of the International Policy Network (IPN), which was formerly the Atlas Economic Research Foundation UK, the UK arm of the US right-wing think tank of the same name. Bate is also an adjunct fellow at the Competitive Enterprise Institute in the USA – one of the think tanks with a history of working against the environmental movement. ESEF and the IEA are part of the IPN, as are many right-wing and free market think tanks across the globe. The IPN in turn is linked to the Sustainable Development Network, a new network of right-wing groups that was formed just before the World Summit on Sustainable Development.

The Sustainable Development Network includes the AgBioWorld Foundation that was instrumental in attacking Ignacio Chapela (see Chapter 8) as well as many of the same free market think tanks in the IPN. The Sustainable Development Network exists to promote a view that goes against the grain of most thinking on environmental issues and is summarized by Julian Morris in his new book. This is namely that 'the world is generally improving and that the rich world in particular has adopted, for the most part, institutions and policies that are sustainable'.[56]

Morris and Bate also network with academics in the USA and the UK who espouse anti-organic and pro-GM views. One of the chapters of the Fearing Food book was co-written by Michael Wilson, who used to work at the SCRI and who now runs Horticulture Research International, and John Hillman who still works at the SCRI.[57] Hillman is on the board of the Bioindustry Association of the UK, whose mission is to encourage and promote biotechnology.[58]

Other academics, too, are linked into this loose network. A prominent contributor on the AgBioWorld website is Anthony Trewavas from the Institute of Cell and Molecular Biology at the University of Edinburgh, who has criticized organic agriculture in the scientific journal, Nature. 'As a plant biologist myself, I have little time for big, insensitive agribusiness' Trewavas wrote in Nature, before saying that 'Going organic worldwide, as Greenpeace wants, would destroy even more wilderness, much of it of marginal agricultural quality'.[59]

Understanding these attacks on organic agriculture is key to the understanding of why apparently independent scientists have taken issue with this form of agriculture. Many of its opponents see the organic movement as standing 'against science', and specifically high-tech science, a significant proportion of which is now funded by agrochemical or biotech companies. 'There is a mindset that is wedded to this high-tech approach and 'scientism', that science is the answer to everything', says Dr Ben Mepham, from the Food Ethics Council. For the FSA, this modus operandi is not to be challenged, but to be embraced.

'What I am suspicious of is that the FSA's starting point begins with the recognition that a huge amount of research in agriculture and food is now commercially driven. We have swapped public science for private/commercial science', concludes Alan Simpson MP. 'The pursuit of knowledge for public or environmental safety has already been ditched in favour of a culture which says we will pursue knowledge for the purpose of commercial gain, and anything that steps in the path will either be excluded or suppressed.'

But just as worrying as the FSA's attack on organics has been its role in 'backing the position of the US government and the biotechnology industry in opposing' strict EU labelling and traceability rules according to GeneWatch UK, which believes that the FSA's position is fundamentally flawed.[60]

Although the FSA maintains that it 'supports consumer choice and recognises that some people will wish to choose not to buy or eat GM foods however carefully they have been assessed for safety', it has been criticized for the stance it has taken on the EU regulations on the Traceability and Labelling of GM food and animal feed.[61] The FSA based its response in part on the recommendations of a report it commissioned on the 'Economic Appraisal of Options for Extension of Legislation on GM Labelling', that was undertaken by National Economic Research Associates.[62] Crucially, however, this report was never peer-reviewed. It was also seen as making assumptions that were biased towards the biotechnology industry.[63] So here was a government agency making a crucial policy decision on science that was not peer-reviewed.

Since 1998 there has been a de facto moratorium on the commercialization of GM crops in the EU. Prior to any commercialization, several EU member states wanted the issues of traceability and labelling resolved to give consumers a choice. Until that point, current regulations only required that food containing GM DNA or protein be labelled. In July 2002, the EU Commission proposed two sets of regulations on GM food and feed and traceability and labelling of GM food and animals feed. The EU was intending to extend its cover on labelling to whole GMOs and animal feeds.[64]

GeneWatch UK outlines how the FSA had 'argued the EU Commission's proposals are expensive and unworkable ... the UK government appears to be more interested in limiting choice and confusing consumers by promoting the concept of a "GM-Free" label in Europe. This "GM-free" label will mean that some products or ingredients will be labelled as GM, some will be labelled as GM-free, but the vast majority will not be labelled even though they may contain GM derivatives. The only information the public would have on the majority of foods would be that: "these products may or may not contain ingredients derived from GMO's"'.[65]

'Disturbingly the UK government and the Food Standards Agency have failed to recognise consumer concerns' says Becky Price of GeneWatch UK. 'They have continually backed the position of the US government and the biotechnology industry in opposing these rules. They tried to propose a GM-Free label which would have left companies that never wanted GM foods having to pay for the privilege of proving their products were not GM.'[66]

The move by the FSA was also condemned by the Consumers' Association who 'remain bitterly disappointed at the anti-consumer stance' taken by the FSA. 'An open and transparent system of labelling, coupled with effective traceability mechanisms, will provide the best basis for consumer choice', said Sue Davies, the Association's Principal Policy Adviser. In contrast to the FSA position, a survey undertaken for the Consumers' Association in the summer of 2002 showed that 94 per cent of consumers think that food containing GM ingredients should be labelled.[67]

While it remains to be seen what rules are finally agreed, without radical reform of both agriculture and science, the situation is set to deteriorate. So what policies could really make a difference?

CHAPTER 11

Towards Safe Food and Public Interest Science

'To put the bounty and the health of our land,
our only commonwealth, into the hands of people
who do not live on it and share its fate
will always be an error.
For whatever determines the fortune of the land
determines also the fortune of the people.
If history teaches anything, it teaches that'

Wendell Berry, American farmer and poet[1]

'You – the customer – hold the fate of farming in your hands'

Anthony Gibson, NFU[2]

Towards safe and sustainable food

When Anthony Gibson, the NFU's Director for the South West and *Farmer's Weekly* Personality of the Year, addressed a meeting on the future of UK agriculture in the autumn of 2002 he said that farmers needed to put their customers first. They needed to create a 'warm glow' with their product. Customers wanted to have a good feeling about what they ate.[3]

There is an increasing consensus that consumers want safe, local, organic fresh food, they want the environment and wildlife to be protected; and farm animals to be treated humanely. They also want to know how their food was produced and whether producers were given a fair price for their products.[4]

Gibson argued that other empowerment measures for farmers were to build their brands, to form cooperatives and deal direct with consumers. One farmer sharing the platform with Gibson has done just that. Guy Watson runs Riverford, one of the country's top organic vegetable farms, which converted to organic in 1988. Recently Riverford expanded to include 11 farms in south Devon to form a cooperative and together they are one of the UK's largest independent growers. Over 5500 households are supplied with veg boxes every week, and there are thriving farm shops.[5]

Despite undertaking all these initiatives, Watson feels there is little future in organic farming if supermarkets continue their vice-like grip on his industry. The reality is that supermarkets are 'holding back the development of organic agriculture', he says, complaining that the market for fresh organic produce is static. 'Whilst organic agriculture is in the hands of the supermarkets, the future is not good', says Watson.[6]

So are supermarkets part of the solution or or part of the problem for agriculture in the UK? Proponents of supermarkets say that they offer round-the-clock convenience, an unprecedented choice and competitive prices. Critics contend that they are not committed to UK farmers and suppliers and do not care about the environment. For UK agriculture to succeed one of two things has to happen: either supermarkets have to massively increase the proportion of local produce they sell, or people have to stop shopping at supermarkets.

Despite the burgeoning local food economy, over 95 per cent of people still do their main shopping at a supermarket.[7] Nearly 75 per cent of our food comes from the 'big five' supermarkets – Sainsbury, Tesco, Walmart-Asda, Waitrose and Safeway.[8] As this book went to press, it was unclear whether Safeway would be taken over by one of the others, but would such a take-over be good for the consumer?

The are lessons to be learned from the USA where Wal-Mart dominates. 'Wal-Mart is not the beginning of competition,' says Al Norman from Sprawl-Busters, who has helped 88 communities fight the company. 'It is the end of competition. Once it has driven out the competitors, it is free to do whatever it wants with its prices.' Supermarkets close down the competition and jobs are lost. UK government research published in 1998, showed that edge-of-town and out-of-town supermarkets have a serious impact on between 13 and 50 per cent of the local market in market towns and centres.[9]

The industry's own figures, from the National Retail Planning Forum report, show that a superstore costs on average 276 local jobs.[10] The Sussex Rural Community Council has predicted that a new supermarket in its region would close all village shops within a 7-mile radius.[11] One of Britain's leading think tanks has calculated that a typical out-of-town supermarket has a subsidy of £25,000 per week over its town centre equivalent, because of pollution and congestion caused by the car culture that out-of-town stores rely on and encourage, but do not pay the effects of.[12]

As supermarkets close down the in-town competition, more and more people are forced to drive to out-of-town stores to buy their food. Between the mid-1980s and the mid-1990s the average distance driven increased by 57 per cent.[13] For the poor who do not have access to a car, this forces them either into excessive journeys by public transport, or to use what shops remain.

Supermarkets are now an oligopoly. When the Competition Commission investigated the supermarkets it concluded there were some 27 practices that were 'against the public interest' in that they 'adversely affected the competitiveness of some of their suppliers', which led to 'lower quality and less consumer choice'.[14]

Everything about supermarkets is on a large scale: production, processing, distribution and consumption are all on an international scale. Guy Watson from Riverford in Devon points out that if he sells his lettuce to Safeway, even though it may be sold in his local supermarket, it travels to Lincolnshire and back to be graded.

The statistics combine to create a damning indictment of why supermarket shopping is not sustainable unless there is a radical rethink of their whole food distribution system. If you buy a traditional meal from a supermarket it has travelled some 24,000 miles. If it is local or bought from a farmers' market, the distance would amount to 66 times fewer food miles. With our food accounting for an ever-increasing proportion of transport usage, it accounts for the emission of more and more carbon dioxide, the leading greenhouse gas. For every calorie of iceberg lettuce we eat, flown in from Los Angeles, we use 127 calories of fuel.[15]

As more and more people shop at supermarkets, the distance our food travels increases. As supermarkets transport produce from one end of the country to another, their share of road freight has rocketed. The distance that food is transported by road increased by 50 per cent between 1978 and 1999 and the food system now accounts for between a 30 and 40 per cent of all UK road freight. Airfreight of food has also expanded significantly.[16]

Supermarkets sell 80 per cent of the fruit and vegetables bought in the UK – a complete turnaround since the 1970s when wholesale markets accounted for 90 per cent of fruit and vegetable sales. The UK was once self-sufficient in fruit and vegetables, now we are only 5 per cent self-sufficient in fruit.[17] As demand diminishes, our orchards and varieties disappear. In the last 30 years half of our pear orchards and over 60 per cent of our apple orchards have been destroyed.[18] With industrial food come the chemicals. Lettuces may look fresh and crispy, but over 11 pesticides are sprayed on lettuces each year.[19]

But it is not just shoppers who are strangled. So too are the suppliers. 'Supermarkets have had a devastating impact on our industry', says Charles Secrett from Thames Valley Growers. 'They have virtually decimated what were rich, varied production areas by taking their business abroad. As long as people continue to buy their fruit and vegetables in supermarkets, I really don't think UK growing has much of a future.'[20] Whilst supermarket support for local produce is growing, it remains woefully inadequate. Waitrose has introduced a charter for small producers.[21] Industry insiders say there is a lot of positive talk, but not much real action.

The government could take steps to regulate the retailing sector and curb the power of supermarkets. There are a number of ways to do this, including:

- Banning, or at least controlling, the proportion of fruit and vegetables sold at supermarkets. Some people have suggested that fruit and vegetable production should be removed totally from supermarket control. 'Supermarkets are inherently incompatible with local food, freshness, with short food miles which we should be concerned about,' says small-holder Alan Beat. Banning supermarkets from selling fruit and vegetables is likely to be deemed political suicide, so in the short term, targets must be set for supermarkets to support local produce. A radical goal would be for 50 per cent of all fruit and vegetables sold at supermarkets within five years to be British grown. This could be introduced by imposing a 'self-sufficiency tax' on supermarkets.
- Storage and seasonality. Some fruit and vegetables are stored for up to a year in nitrogen. Storage limits could be reduced to a month, so that supermarkets are forced to sell more seasonal produce, not all-year-round tasteless and uniform produce.
- A reduction in pesticide levels. Targets should be set to reduce pesticide levels in all fruit and vegetables in the UK.
- A ban on opening up any more out-of-town developments. If there is to be expansion it has to be in town and on a smaller scale.
- Reduce transportation. Aircraft fuel has to be taxed. This would reduce the ridiculous 'great food swap' where we import and export the same product. For example, in 2000 we exported some 210,000 tonnes of pork and bacon at the same time as importing 520,000 tonnes.[22]
- Introduce an ecological label scheme that shows the shopper the environmental impact of distribution and the food miles involved in the product.

If these measures do not kick-start and sustain local food production, then subsidies for local food production may be a politically acceptable idea within the context of a comprehensive action plan with targets for sustainable food production systems.

Common Agricultural Policy (CAP) reform

Anthony Gibson of the NFU talks about three big levers that affect farmers, that are basically outside their control. The first is CAP, the second is exchange rates and the third are world markets.[23] Everyone agrees that reform of CAP is coming, but it remains a thorny issue with European politicians. CAP will eventually go, concluded the Policy Commission on the Future of Farming and Food, but real reform might take time. One major recommendation for reform was 'modulation', where subsidies would be changed from food production to environmental and rural development schemes. 'The UK currently plans to modulate up to 4.5 per cent of direct payments by 2006,' said the Commission, urging quicker reform of 10 per cent modulation from 2004.[24]

Reform, when it does come, may not be comprehensive enough or even on the right track. One of the ideas is 'de-coupling', where all the bureaucratic paperwork is bundled into one payment. Whilst farmers would welcome any reduction, the payment would be made regardless of what the farmer does on the farm, and may even be to the landowner rather than farmer. This could encourage people to buy land purely to receive subsidies.[25]

Indeed reform is unlikely to go as far as many want. A coalition of 14 major development and environment groups argue that reform must include CAP being transformed to incoporate new 'social, environmental, animal welfare, rural and international development, and health objectives'.[26]

The problem for the UK's farmers is that the more they produce high welfare, sustainable food, the more they are vulnerable to cheap, low welfare imports. 'We can't compete', says Gibson, 'farmers are crying out to stop sub-standard imports'. Gibson argues that because import controls are politically unacceptable, the only way option for farmers is to build local brands that people want – part of the 'warm glow' strategy.[27] But are import controls really so unacceptable?

Import controls

Opponents of import controls say they break the rules of free trade and leave our exports open to retaliatory action. But we already subsidize our farmers to the tune of £3 billion a year, so there already is a market distortion. Whilst CAP is widely regarded as a system that does not provide what people want, it is time to produce a system that does. It is time to think the unthinkable. Two leading experts on international trade and food policy are Colin Hines, the international Green Party advisor, and Indian activist, Vandana Shiva, who was named by *Time Magazine* as one of its 'environmental heroes' in its special Earth Summit edition in 2002.

Hines and Shiva argue that 'the major shortcoming of the Curry Report [on the Future of Farnming and Food, see Chapter 4] – and of the CAP reform options under consideration – is the failure to grapple with the realities of ever increasing trade liberalization and ever growing competition in global food markets. It is simply not credible for farmers to be asked to both raise their environmental and animal welfare standards, and at the same time to be ever more internationally competitive in global markets. If higher standards are the goal, then countries will have to be prepared to protect their standards against the international trade from cheaper imports from countries that do not meet those standards'.[28]

Others agree: 'The most vital step is to protect domestic food production from competition from imports', argues Pippa Wood from the Family Farmers' Association, with Michael Hart from the Small and Family Farms Alliance. 'Nearly all the food we produce can be produced more cheaply elsewhere, as other countries have less rigorous regulations on welfare, hygiene and environment. Most of them also have cheaper land and lower wages; some subsidize production more heavily than we do. British farming has had many extra problems to contend with in recent years, but the WTO rules insisting that we accept unlimited imports, regardless of their production methods, look set to be the "coup de grace" which will finish us all.'[29]

Their message sticks with small-holders and small farmers. 'We need to change the legislative framework within which we work,' says Alan Beat, 'so that it is financially possible for a farmer to farm in more traditional environmentally friendly ways and still earn a living, rather than be propelled by sheer market forces in entirely the wrong direction'.

Unless there is a change in economic direction there will be no ethical and sustainable agriculture. 'Unfortunately, the domination of contemporary thought by the language of high-tech and global market economics has led to the ethical perspectives becoming highly marginalized', laments the Food Ethics Council. The Council argue that this preoccupation is misguided and argues that 'instead of relying on a simplistic cost/benefit approach, policies should take account of widely accepted ethical principles, which, crucially, also place value on rights and fairness. Among many other advantages, adoption of the latter approach in the past would have prevented disease outbreaks such as

BSE and FMD, and not only saved many lives but also many millions of pounds'.[30]

A number of recommendations may be made, including:

- Import controls – it is time for the 'gradual introduction of import controls to protect those goods that can be produced domestically from imports that could otherwise threaten the rediversification of national agricultural systems'.[31]
- Tighter border security – stricter border controls would stop the flood of cheap and illegal meat that was seen as being responsible for foot and mouth disease.

A true democratic voice for farmers that promotes local family farms

Small farms have consistently been shown to be more productive per acre, be less polluting, and better for employment, wildlife and diversity. They are likely to have more hedgerows and more deciduous woodland. But they are unable to compete with the powerful forces of global trade and so many have gone bankrupt. In 1939 there were some 500,000 small farms in the UK, today there are just 168,000.[32]

Many farmers moan about the policies and lack of democracy of the NFU and the fact that it looks after the big farms not small ones. Sentiments like these lead to the formation, in November 2002, of a new alliance called FARM that argued that over two-thirds of farmers want a new organization to represent them.[33]

It is the NFU's conservative resistance to change that makes many farmers despair. 'They do not understand what the environment means', says Mike Downham, who farms 220 acres in north Cumbria. 'They don't have any concept of biodiversity or pollution and like so many unions, they're just determined to hang on to the subsidy system which has kept them going for years.'[34]

Indeed the NFU likes to call itself the 'democratic organization representing farmers and growers in England and Wales'. For the NFU to be truly democratic it has to take on board the voices of the 80,000 small-holders and small farmers who are 'Countryside' members of the Union. All members of the NFU, including the Countryside members, should be allowed a vote. If the NFU were truly democratic, sustainable change in the UK countryside might happen at a quicker pace.

Pay farmers a decent price for their product

Whilst politicians will find the concept of import controls hard to swallow, consumers may equally baulk at the thought of paying more for food at the shops, even though they pay indirectly far much more than they realize due to subsidies. Hart and Wood maintain that 'it is low farm gate prices' that have fuelled much of the intensification of agriculture. As much as price is part of the problem, it is also part of the solution. In the 1960s, consumers spent 30 per cent of their income on food. In 2001 it was just 10 per cent.[35] Fifty years ago for every pound we spent in the shops on food and drink, some 50–60p went to the farmer. Today the figure is a lowly 9p.[36]

'It is up to society to pay a price for the product they buy', argues Alan Beat. 'It is going to be an enormous educational process. Gradually we need to change people's perception of what cheap food actually means, and how local more expensive food is actually better for them in the long-run.' Beat argues that it is necessary to 'reorganise the retail price index, so the inevitable increase in the price of food would not be seen as an automatic inflation. I think the retail price index is a disaster to better more expensive food'.

Many with extensive knowledge of farming argue that it is inevitable that people will have to pay more for food. Peter Stephenson from Compassion in World Farming argues that 'we can afford our holidays abroad and our computers and the lottery, and the idea that we as a society cannot afford better food is not true. We are responsible for the state that farm animals live in and we must realise its going to cost us a little bit more'.

Part of the way to persuade consumers to pay more is through education, argues James Coleman, a free range duck producer from Devon, who was Young Farmer of the Year in 2002. 'I believe', says Coleman, 'that if the consumers were in a position where they knew the difference between a British product and a cheaper imported product, a considerable number would turn towards the British product'. Coleman believes that only food reared in Britain should be able to call itself British product, not food that is simply packaged here.[37]

Recommendations to address these issues include:

- Remove food items (currently around 10 per cent) from the retail price index.
- The government should fund an education campaign to promote British food and, more importantly, local food. This leads on to the next point.

Promote local food

Whilst all of the above should be encouraged, radical reform is not easy. On the surface there is a renascence of local food initiatives that are happening all across the country such as Community Supported Agriculture, consumer cooperatives, producer cooperatives, box schemes, growing your own – gardens, allotments, and community gardens – local shops, farm shops, 'pick your own' and farmers' markets.

Two of these initiatives enjoying great success are box schemes and farmers' markets. There are now some 300 organic box schemes registered with the Soil Association. Box schemes represented £22 million in sales in the period 1998/1999, delivering to 45,000 families per week in the UK. It is now over five years since the first farmers' market opened in Bath. Since then, demand has rocketed. There are now some 15 million visits to the 500 plus farmers' markets across the country, worth an estimated £166 million a year. The demand for local produce is growing so much that producers are finding it difficult to keep up. 'There are several threats', says James Parvitt from the National Association of Farmers' Markets, 'including the fact there are not enough farmers and growers to match the number of markets'.

But as farmers' markets grow, there is increasing pressure to relax the rules, especially over what 'local' means. 'The term "local" is changeable', says Parvitt. The large producers are not happy either. The Food and Drink Federation, has tried to argue that local produced food entails higher spoilage.[38]

Such is local food's revival, it is seen by the Curry Commission and supermarkets 'as the next major development in food retailing'. There is a growing realization that more local food should be produced within the political establishment. Environment Minister, Michael Meacher, has called for food that is 'more localized, less internationalized, less dependent on chemical fertilisers more low impact, and more organic'.[39] 'What we need to develop is the idea that we are going to produce safe, sustainable and ethically produced food within localities which actually sell into local markets', says Colin Breed, from the Liberal Democrats. Breed has endorsed the *Western Morning News*' 'Buy Local' campaign that encouraged businesses and consumers across the southwest to use local food. Businesses and food celebrities such as Hugh Fearnley-Whittingstall as well as the Prince of Wales have backed the campaign.[40] At the Royal Agricultural Show in the summer of 2002, the Prince called on people to change their shopping habits by stopping their demands for cheap food. He also criticized the government for not buying local food for hospitals, schools and the military.[41]

The NFU have pointed out that whilst the last couple of years has seen a much greater willingness to buy local, current EU procurement rules make it illegal for local firms to be given catering contracts. The current focus is all on price.[42]

Many of the other recommendations outlined above and below will benefit the supply of local food, but the following may also be considered:

- EU rules should be changed to actually encourage local firms to provide local food for hospitals and schools.
- Businesses and caterers should be given tax-incentives to buy local food.
- Both local and national governments should set specific targets to increase the production and consumption of local food.

High animal welfare standards in both production and transport

Although British farms have a better animal welfare record than many, there is still a long way to go. In the UK, some 750 million broiler chickens are reared a year – 98 per cent of them intensively. Their last week is spent on the size of an A4 piece of paper.[43] A quarter of the 32 million laying poultry suffer bone fractures.[44] A recent survey found that 20 per cent of chicken meat tested and 10 per cent of eggs contained residues of drugs 'deemed too dangerous for use in human medicine'.[45] The stress put on dairy cows has been likened to a 'man jogging for 6–8 hours per day, every day'.[46]

Conventional pig farming is criticized too; 13–14 million pigs are eaten in the UK every year. These are produced by some 700,000 rearing sows. 'The majority of pigs continue to be ruthlessly factory farmed,' says Peter Stephenson, the Director of Compassion in World Farming.

Stevenson highlights major areas of concern, such as overcrowding of fattening pigs, which are kept indoors all their lives, in barren environments and the docking of tails to stop pigs chewing the tails of other pigs. 'They will be given no straw,' he says. Most conventional pigs receive antibiotics leading to concerns about antibiotic resistance. Others, too, argue that there has to be a change in the way animals are reared. 'If there is any moral justification for eating meat at all, then there is a kind of contract', says celebrity chef Hugh Fearnley-Whittingstall. 'We'll eat you but until we do so we'll give you the best life possible. That has to be the deal.'[47]

Fearnley-Whittingstall sources some of his meat from Ian and Denise Bell's 68-acre farm on the Dorset coast, which is owned by the National Trust. Whilst many farmers are in crisis, the Bell's are making a living using high welfare and sustainable practices, encompassing the biodynamic end of organic production. 'We have more customers than we have produce,' says Denise. 'We have proved over the last six years that you can live without agrochemicals and pharmaceuticals,' adds Ian. To those who question their unique methods, Denise says 'we have the purest meat in England with the finest eating quality, with consistency'. Other celebrity chefs rave about their products including Michel Roux, Raymond Blanc and Nigella Lawson who argues that 'their pork is so much better than any pork you have ever tasted or could ever hope to taste'.

The farm is deliberately small-scale, slow and extensive. Whereas in conventional pig farming the piglets are removed from their mothers after about 3–4 weeks, the Bell's piglets are allowed to wean naturally for up to five months, giving them natural immunity from their mothers. By conventional standards the pigs are fed a gourmet healthy diet that includes fresh fruit and vegetables. There is no need for tail docking, and there is an abundance of straw. The piglets are kept on the farm for a year and are then taken straight to the local abattoir. This contrasts with conventional agriculture, where animals often travel hundreds of miles to market or to the slaughterhouse, an issue that the Curry Commission said should be rectified as a 'high priority'.[48]

Ian and Denise will not stock supermarkets, preferring to sell to loyal customers, restaurants and even operating a growing mail order business. 'We think the supermarkets are largely to blame, they are constantly telling people that they want cheap food', argues Denise.

But there is a catch, which is the cost, and the periodic supply, which means regular customers have to buy a freezer. Their pork, sold under the Heritage Prime label is about £4.95 per pound, which is more expensive than in the supermarket. But the Bells argue that the extra cost has not gone to a middleman or a supermarket, but has gone into making a happy pig. They also see a trend in people being prepared to pay more for their food. 'In the last three years things have changed dramatically, because people are so concerned after BSE and FMD', says Denise. 'Food is becoming high on the list of priorities. We have people phoning up saying that they are not going on holiday this year as they want to spend that little bit extra on decent meat.'

Others, too, are finding the benefits of selling traceable, high quality meat direct to the consumer and working cooperatively. Wealden Farmers Network is a collection of five livestock farms in East Sussex that are working together to produce, prepare and supply locally grown, fully traceable meat to local people.[49]

Concentrating on using local breeds, between them the farms have about 50 suckler cows and 400 breeding ewes. Although only one of the five farms is certified organic, all farms use organic principles. 'We feel that the success of the network is due to its small size, to our direct marketing strategy to local customers at reasonable prices, and to the very high environmental and animal welfare standards that we apply', says network member Simon Bishop.[50]

Many farmers, like those in the Wealden Farmers Network and the Bells are farming in a sustainable, humane way. Whilst many consumers still cannot afford to pay extra for their food, these examples show that there is an alternative to cheap, industrial agriculture. In addition, if supermarkets were made to pay the full ecological cost of transporting their produce, these farmers would suddenly be on a level playing field.

Other recommendations which could enable farmers to produce food in the ways discussed above include:

- A radical proposal is to remove meat from supermarket control. 'Supermarkets just shouldn't be allowed to sell meat', argues Ian Bell 'that would be a great way forward. The morals, the ethics, it doesn't suit supermarket production'.
- Stop the live export and long distance movement of animals. Not only is this long distance movement cruel, but through it you risk another major outbreak of foot and mouth disease. The RSPCA has warned of a 'ticking time-bomb' of a new outbreak of foot and mouth unless action is taken to 'limit the distance or frequency of journeys during an animal's lifetime'.[51]
- Reintroduce local abattoirs with strict welfare and hygiene conditions, that are subsidized by business incentives and tax cuts.
- Subsidize extensive and high welfare dairy herds and meat production.
- According to Compassion in World Farming existing farm assurance schemes, like the well-known Little Red Tractor, actually do 'very little' for animal welfare.[52] A new animal welfare labelling scheme could be introduced that complements the ecological scheme and gives shoppers information on true welfare standards and the distance travelled to slaughter. This could also include the Cattle Passport scheme so consumers know where the cow has come from, and there is traceability back to the farm.

Environmentally beneficial or benign in its production.

There is no doubt that organic agriculture is enjoying a period of influence and expansion unknown in its history. Sales of organic food are growing at record levels. The market topped £600m in 1999–2000 and is expected to be worth £1 billion within two years.[53] In its latest report, the Soil Association notes that its members now stand at nearly 20,000 – its highest ever level.[54] The Curry Commission was labelled by the Soil Association as the 'biggest breakthrough in the organic movement's history'.[55] But whilst the government responded in July 2002 to the Curry Commission and the growing demand for UK agriculture, by announcing a target for British organic producers to supply 70 per cent of the domestic market, environmental groups criticized the lack of any firm timetable for achieving the target.[56]

However, there are positive signs; the amount of organic land in the UK almost doubled in 2002. But despite rocketing demand for organic produce in the UK and UK suppliers often being able to meet this demand, supermarkets source much of their organic produce abroad. Overall, 65 per cent of organic produce sold in the UK is imported, with Tesco the worst performing supermarket with 80 per cent of its organic produce coming from overseas. The best retailer is M&S, which buys 85 per cent of its organic produce in the UK.[57] Some supermarkets are trying to buck the trend: Sainsbury's, named by the Soil Association as organic retailer of the year, has pledged to reduce its organic imports to 45 per cent by 2004.[58]

But supermarkets fail on price. Research by Dr Anna Ross from the University of the West of England found that 'supermarkets are the most expensive of all organic food retailers and they have the smallest range of fresh produce'. Dr Ross concluded that 'organic consumers would get a better deal and more choice if they switched from buying organic vegetables and meat from the supermarkets to independent and direct sellers'. Ross found that supermarket vegetables were 78 per cent more expensive than the vegetable box from Riverford in Devon. Over the past two years the difference in price between supermarket and box scheme vegetables had increased by 39 per cent.[59]

There are already positive moves to help organic and/or local sustainable food producers. In part to help consumers identify British, English, Scottish, Welsh and Northern Irish organic food, a new set of logos and standards was developed by the Soil Association in the autumn of 2002. The move was intended to help shoppers pick out national and local food in shops and enable organic farmers and food manufacturers to clearly indicate to customers the origin of foods.[60]

Another initiative is the formation of the Wholesome Food Alliance (WFA), that hopes to offer farmers an alternative if they cannot afford the costs of organic certification and the Soil Association's closeness to supermarkets. 'The calls I get usually complain about the cost of Soil Association certification', says Director Phil Chandler. 'For the smaller grower it is not feasible to cough up nearly £500 for the privilege of using the Soil Association logo. There is also a lot of unrest about their association with supermarkets. People don't like the fact that they are so intertwined.'

The WFA hopes to attract small growers who sell locally, and already has a mailing list of several hundred. 'We are supporting the local and smaller grower,' says Chandler. 'We insist that our members don't use synthetic fertilizers, herbicides or pesticides and we are offering the logo to farmers and smallholders who want to sell in their area. It is local and traceable.'

A number of recommendations could encourage organic growers:

- The government should adopt a strict timetable for the adoption of new organic targets.
- Supermarkets should either set or be set strict targets for selling organic food, grown in the UK.
- There should be a timetable for the reduction of pesticide residues in food, across the board.

Healthy food

'One of the worst legacies of the BSE crisis is that people now see food as a vessel of disease, not as a source of health', says Dr Tim Holt, a critic of the government's handling of the BSE crisis. The UK is among the worst affected countries for diet-related ill health, which includes cardiovascular disease, cancers (eg breast, colon), strokes, diabetes and hypertension.

Instead of things improving, they are likely to get worse. Consider advertising and children. Current voluntary codes of advertising practice fail to protect children from commercial advertising that promotes fatty, sugary and salty foods, directly to children. Over 80 national public interest organizations have called for legislation to ban this type of advertising to children.[61] The organization behind these proposals, Sustain, notes its disappointment at how 'the FSA seems unwilling to pursue its commitment to develop a code of practice or guidelines on advertising to children, until their research review update is complete'.[62]

A report edited by Professor Tim Lang from the Centre for Food Policy, Thames Valley University and Dr Geof Rayner, Chair, UK Public Health Association, highlighted the intricate link between food farming and health. The report argues that 'health, therefore, is the key to the future of farming and food in England'.[63] The report called for the following proposals that should all be encouraged:

- 'A new integrated approach to health – linking nutrition, food safety and sustainable food supply, an approach here called ecological public health – would provide the food and farming industries with a new framework that they desperately need.
- Farming practices that sustain environmental improvement – namely to reduce energy use, to increase variety and biodiversity and to conserve amenity for the public – are congruent at the same time with those that are most effective in promoting public health.

- There should be a population approach to farming and food policy on top of the current market and individual focus.
- The food supply chain needs to deliver affordable, health-enhancing and accessible diets for all, not just for those who can afford it.'[64]

Lang and Rayner noted that there has to be a reduction of diet-related inequalities to tackle social exclusion and poor access that should be at the heart of farming and food systems.

Another radical idea would be to make cooking a compulsory part of the National Curriculum, and ensure it celebrates local and seasonal food. All schools should be encouraged to have an allotment or vegetable garden.

If the government committed the same amount of time, energy and resources into promoting a healthy local diet with good nutritious food available for all, then things would be very different. Instead they are preoccupied with promoting GM.

GM

There needs to be a change in the terms and conditions of any commercialization of GM, including a different set of values.

Consumers not companies come first

Consumer choice has to be guaranteed at no extra cost. It is not acceptable to force people to eat GM food. We know the majority of Europeans do not want to eat GM. We know that an increasing percentage want to eat either organic or local produce, and ideally local, organic food. The right of the consumer to have this choice has to be paramount. If GM commercialization goes ahead, choice must be legally binding and guaranteed at no extra cost. All GM products, including part or whole ingredients and derivatives, have to be labelled. This includes animals bred on GM feed.

Legal liability

The biotech industry has to be legally liable for the economic, environmental and health implications of their products. There cannot be commercialization without this legal framework.

Scientific certainty

The notion of substantial equivalence is now so discredited that this cannot be used as a benchmark for GM approval. Systems must be developed to examine GM foods for unintended changes to the food, not just to look at the effect of the intended changes as happens now. As well as more sophisticated chemical tests, improved biological testing is needed.

But this call for more safety testing should not be allowed to lead to more animal testing. An earlier evaluation must decide whether the GM crop is necessary and relevant to a new, sustainable agricultural system. Only if it is clear that there is a need for the GM food should animals be used.

No genetic pollution

There can be no genetic harm to plants, animals or people and no GM contamination of wild species, food crops or seeds. GM gene containment has to be the way forward, although technical fixes such as Terminator Technology, where seed is sterile and cannot be used by farmers, are not acceptable solutions. Domestic crops should not be contaminated – systems must be put in place to protect non-GM and organic farmers.

GM crops that have wild relatives close by should not be grown unless it is certain they cannot cross-pollinate. In northern Europe, this means no GM oilseed rape or sugar beet; in south America, it will mean no GM maize.

No patents on GM resources

Patents are being granted to companies and individuals on genes, seeds and plants, redefining life as an invention of industry and turning over the centuries-old view of genetic resources being the heritage of mankind. Patenting has driven the consolidation of the biotechnology industry and facilitated the take-over of local seed producers by multinational corporations. A reversal should take place and patents should be banned.

Ban the cloning and genetic modification of animals

GM and cloning of animals should be banned.

Public interest science

We stand at a crossroads for UK science too, as well as agriculture. But just as agriculture is crying out for reform, so too is science. Although science is only part of the policy-making equation on food, scientists have increasingly been used by politicians to justify their political action. We know already that many, many scientists remain deeply sceptical about the direction of science and its corporatization and politicization. They resent having to get the begging bowl out and walk cap in hand to companies. Science has become a battle between precaution and exploitation.

Do we adopt the high-tech, biotech future, where the sole function of science is as a vehicle of wealth creation, where dissent is not tolerated, where diversity of ideas and engagement of intellect are memories of a bygone age? In this future, science has become subservient to its political and corporate masters. Scientists have been placed in an impossible position trying to reassure an increasingly sceptical public that the ever-expanding array of GM foods and cloned animals are both safe and a necessary part of a diet, that every day lacks more fresh, natural or even local produce.

Scientists should not allow themselves to become pawns of manipulation in a political game, just as they have allowed themselves to become an integral part of the government's plans for wealth generation. Corporate links have become the norm, rather than the exception.

Towards public interest science

Instead of monolithic corporate science, where critical voices are suppressed, there needs to be a radical rethinking of the underlying principles that underpin the new scientific doctrine, according to Dr Andy Stirling, from the Science Policy Research Centre at Sussex University. These are simply:[65]

- 'Humility – by acknowledging the limits of scientific methods and models and the intrinsic subjectivity of the assumptions that always frame the science.
- Pluralism – by engaging with different interest groups in society to ensure that the right questions are asked of the science and that the interpretations are socially robust.
- Diversity – in that science advice should not just be about making definitive prescriptions, but should convey a range of different recommendations, including those from dissenting voices, so that the final political decisions are made more transparent and accountable.'

Rebuild public interest science

There need to be fewer industry advisors on the councils, committees and boards of research institutions and universities and more lay people. Society has to believe that science is being undertaken for their benefit, not at a cost to them. If public science is being driven by private greed, people will never trust it. People feel that 'if we can never fully know the consequences, then we had better at least ensure that the purposes driving the enterprise, and the interests which control the responses to the resultant surprises, are good ones'.[66] Science and scientific advice surrounding GM has to be in the public, not corporate interest. Public science is totally different to a scientific dogma that pursues wealth creation.

Part of this approach would include the following suggestions.

Public body on technology

The public needs to be involved in debating what science can and cannot do and what the public expects from scientists. This was recognized by the Policy Commission on the Future of Food and Farming, called the Curry Commission, which said 'consumers need to be involved both in the framing of debate on science and technology, and on the bodies dictating policy where it affects them'. The commission recommended 'that the government should set up a new "priorities board" for strategic research, involving government, academic, consumer, environmental and industry representatives to set the agenda for public research on farming and food matters'.[67]

The government should establish a wholly independent commission or academy, which analyses technological developments. This would involve a diverse range of people – not just scientists – and would look at many issues, such as the ethics of and need for technological development. It would not be acceptable for government or industry to introduce profoundly different technologies without them being scrutinized by this public body first. Therefore, there needs to be a publicly funded committee that looks at technological innovation and whether there is an actual need within society for that technology.

Realign Foresight

Science has to be free of political coercion and corporate control – it should have more democratic control and meet the needs of a wider range of society, not just big business. The corporate alliance of science has to change. The government should realign the Foresight process to move away from the emphasis on competitiveness to one of precaution and diversity of ideas.

Capping of corporate funding

The percentage of corporate funding of university and individual departments should be capped at no more than 10 per cent. To take up the deficit in funding, there would be a massive increase in public scientific spending. For the economists who baulk at this idea, it is worth remembering that the combined cost of BSE and foot and mouth was in the region of £16–20 billion.

Both of these disasters would have been considerably smaller, or may even have been prevented, with the proper public support and finance of key laboratories and scientists. Where is the logic of a relationship between the government lab at Pirbright and a Dutch company that stops it working with the Americans at a time of crisis on foot and mouth and precludes the utilization of a solution that could have saved the lives of millions of animals?

Once celebrated corporate links, such as the John Innes Centre and Syngenta, have recently backfired, for example when Syngenta pulled out in September 2002. This announcement totally undermines the government's strategy for the commercialization of our research centres. If a big company like Syngenta could not make it work then many will ask, who can?

The reality is that basic research needs would be funded by the public purse. In the long run this would create true innovation of ideas, as scientists would be free to be more creative and inquiring, not just worrying what their funders will think

There should also be a massive increase in scientific research into more ecologically benign technologies in the areas of food and alternatives to GM, especially organic agriculture.

A radical rethink on scientific advice given to select committees and how scientific committees are chosen and run

The more the politicians rely on scientific advice the more it becomes fundamentally important who that advisor is and the advice they give. Once again, there needs to be less industry on these committees and more consumer and lay representation, making up at least half the committee. If there is only one token consumer representative on a committee, any majority voting system would simply overrule them.

One leading scientist said simply 'we don't know how to choose expertise. Unless that question is thrust at all democracies that is their biggest, weakest link'. The BSE and GM sagas are littered with examples of parliamentary committees that do not understand the issues criticizing key scientists.

The whole scientific committee structure has to change to be totally independent of government and Whitehall. So many scientists have moaned at the rigging of the scientific committee system, and the failures of BSE and foot and mouth disease underline this massive failing. Committee members cannot be chosen by Whitehall mandarins who are only picking scientists whom they believe will protect the government position, but give the veneer of independence. Committee minutes cannot be altered by civil servants and committee members cannot be pressurized into adopting positions that they do not want to adopt.

Whilst bodies such as the AEBC have broken the mould in recent years and its formation has to be welcomed, even its members feel that much, much more could be done. Science is still seen as a dominant ethos and other social, ethical or economic concerns are viewed as less important or rational. The way in which science has been shaped, and the values it conceals, have to be recognized if progress is to be made.

Scientific committees must also provide Ministers with a range of pluralistic options and recommendations looking at different aspects of science and technology. When they give their advice, they must have looked at all the social and economic factors influencing the outcomes. For example, when scientists in the BSE saga said that beef was safe they did not visit slaughterhouses to see that the measures they proposed were impractical, and so infection routes continued for many years.

Some even believe that the whole scientific advice system, in its current form, should be scrapped. Professor Richard Lacey believes that a combination of factors makes current scientific committees 'virtually untenable', especially in relation to GM, due to the potential explosion in numbers of new products, and just the vast amount of material that the experts need to process.

Democratize The Royal Society and other leading learned societies

If The Royal Society and other learned societies are to remain at the forefront of British science they should be much more democratic, transparent and open institutions, inviting many more members who are female, younger or of other ethnicities.

The precautionary principle

Knowledge creation, not wealth creation has to be the goal. If we have learned anything from BSE and foot and mouth, it is that the precautionary principle needs to be the overriding factor that guides science, not commercialization.

The precautionary principle states that action should be taken before there is strong evidence of harm, particularly if there is a possibility that that harm will be irreversible and delayed. It is summed up by the popular phrase: 'Better safe than sorry'. It 'has nothing to do with anti-science, and everything to do with the rejection of reductionist, closed and arbitrarily narrow science in favour of sounder, more rigorous and more robust science'.[68] This must include risk assessments that in the past have been proved to be widely wrong and the safety testing of GM foods. The safety testing up until now has been neither rigorous nor robust nor comprehensive. The only reason to rush GM on to the market is the commercial one. The Expert Panel on Biotechnology from the Royal Society of Canada argued that the tenet of the 'precautionary principle should be respected in the management of the risk associated with food biotechnology'. They say that the central force of the precautionary principle is very simple. 'Given the potential of at least certain kinds and magnitudes of harms, reasonable prudence would slow the development of technologies pending stronger assurances of their safety or the implementation of active measures to guarantee safety.'[69]

A guarantee of safety may be hard, but is the only way forward for politicians, the scientists and the farming industry, including the GM companies. After everything that has happened to UK agriculture over the last 20 years, that guarantee may be the only hope of restoring people's trust in the food they eat. The public has been betrayed over BSE, they have been betrayed over foot and mouth disease, they cannot be betrayed over GM.

Notes

All quotes are from interviews with the author, unless otherwise stated.

Chapter 1 Introduction

1   Blair, T (2002) 'Science Matters', Speech to The Royal Society, London, 23 May.
2   Taken from George Orwell's essay *Politics and the English Language*, published in 1946.
3   Bonino, E (1998) 'Forum – Consumers' Expectations and Food Safety: New Foods – What do we need? What do we want?', Vienna, 12 October.
4   Taylor Nelson Sofres Consumer (2003) *Consumer Attitudes to Food Standards*, A report for the Food Standards Agency, February.
5   Milburn, A (2003) *Letter to Lester Firkins of the Human BSE Foundation*, Department of Health, 7 January.
6   GeneWatch UK (2002) 'GeneWatch UK Welcomes MEPs' Vote to Give Consumers Choice About GM Food', Press Release, 3 July.
7   Consumers' Association (2002) *GM Dilemmas – Consumers and Genetically Modified Foods*, Policy Report, London, September, p7.
8   Ibid.
9   Friends of the Earth, (2001) *Response to the Policy Commission on the Future of Farming and Food*, London, October, p9, 10; Vidal, J (2001) 'Global Trade Forces Exodus from Land', *The Guardian*, 28 February, p4.
10 Royal Society of Canada (2001) *Expert Panel on Biotechnology*, p206.
11 Vidal (2001) op cit, p4; Ackerman, J (2002) 'How Safe?' *National Geographic Magazine*, May, p19.
12 Gibson, A (2002) *The Future of Farming in Devon*, Talk at the Barn Theatre, Dartington, Devon, 26 October.
13 National Farmers' Union (2002) *UK Agricultural Review – Farming in Crisis*, London, June; Gibson (2002), op cit.
14 Ibid.
15 Hawkins, K (2002) Speaking During the Q&A Panel Session, at the Over Thirty Months Rule Review, Public Meeting Organized by the Food Standards Agency, Congress Centre, London, 2 July.

16   Friends of the Earth (2001) op cit, p5–6.
17   Lawrence, F (2001) 'Hidden Costs Behind the Checkout Prices', *The Guardian*, London, 28 February, p5.
18   Policy Commission on the Future of Farming and Food (2002) *Farming & Food – A Sustainable Future*, January, p67.
19   Friends of the Earth (2001), op cit, p6.
20   Schlosser, E (2001) *Fast Food Nation – What the All-American Meal is Doing to the World*, Allen Lane, Penguin, pp53, 54, 115, 121.
21   *The Guardian* (2001) 'From Farm to Plate – A Sick Industry,' London, 28 February, p5.
22   Boseley, S (2002) 'Food Industry Blamed for Surge in Obesity', *The Guardian*, London, 13 September, p10; *The Guardian* (2002) 'Gross Negligence', Editorial, London, 13 September, p 19.
23   Henery, M (2002) 'Children Spending £8M on Junk Foods', *The Times*, London, 17 September, p4; Food Ethics Council (2001) *After FMD: Aiming for a Values Driven Agriculture*, Southwell, p15.
24   Food Ethics Council (2001), op cit, p17.
25   Revill, J (2002) 'NHS Wakes Up to Child Obesity Crisis', *The Observer*, London, 3 November, p5.
26   Blythmann, J (2002) 'Strange Fruit', *The Guardian*, Weekend Section, 7 September, pp20–24.
27   Ackerman (2002), op cit.
28   Weale, S (2002) 'The Great Sell-by Con', *The Guardian*, London, 7 November, G2, p10–11.
29   Davidsottir, S (2001) 'Farm of the Future', *The Guardian*, London, 22 August quoted by Food Ethics Council (2001), op cit, p27.
30   Lean, G (2002) 'American Cloned Food on its Way', *The Independent on Sunday*, London, 6 October.
31   Burke, Professor D (1999) 'No Big Deal', *The Guardian*, London, 13 February, p21.
32   International Service for the Acquisition of Agri-biotech Applications (2003) *2002 Global GM Crop Area Continues to Grow for the Sixth Consecutive Year at a Sustained Rate of More than 10%*, Press Release, 16 January.
33   Shubert, R (2002) 'Pushin' Roundup via Roundup Ready wheat?', *CropChoice News*, 17 June; www.cropchoice.com
34   Ibid.
35   Hatchard, G (2002) 'US Farmers Reap Heavy Penalty for Sowing GM Crops', *New Zealand Herald*, 27 August.
36   Soil Association (2002) *GM Crops are Economic Disaster Shows New Report*, Press Release, Bristol, 17 September.
37   Fernandez-Cornejo, J and McBride, W (2002) *Adoption of Bioengineered Crops*, Economic Research Service Agricultural Economic Report No. AER810, USDA, Washington, May, p24.
38   Fox, B (2002) 'GM Plants no Panacea', *New Scientist*, London, Vol 175, 17 August, p22.
39   Elias, P (2002) 'Biotech Lobbyists' Clout Grows in Washington', *The Associated Press*, 2 June.

40   *Farmers Weekly* (2002) 'EU Threatens Biotechnology Development', Sutton, 19 April, p21.
41   Niss, J (2002) 'Bush Baits Brussels Over GM Crops', *The Independent*, London, 25 August; *Washington Post* (2002) 'Europe's Biotech Madness', Editorial, 28 October, pA18.
42   Vidal, J (2002) 'US Presses Africa to Take GM Foods – Six Countries Beset by Starvation Reject Offer of Altered Crops', *The Guardian*, London, 30 August.
43   Pearce, F (2002) 'UN is Slipping Modified Food into Aid', *New Scientist*, London, 19 September.
44   Harrison, D (2002) 'Scientists Fight Back Against the Tide of Green Lobby Claims', *The Sunday Telegraph*, London, 1 September.
45   Greenpeace (2002) *USAID and GM Food Aid*, London, October, p1.
46   Ibid.
47   Burson-Masteller (1997) *Communications Programmes For EUROPABIO*, January.
48   Phillips, Lord, Bridgeman, J and Ferguson-Smith, Professor M (2000) *The BSE Inquiry Report. Evidence and Supporting Papers of the Inquiry into the Emergence and Identification of Bovine Spongiform Encephalopathy (BSE) and Variant Creutzfeldt-Jakob Disease (vCJD) and the Action Taken in Response to it up to 20 March 1996*, The Stationery Office, London, October, Vol 1, pp20, 227.

Chapter 2 Safe to Eat

1   Phillips, Lord, Bridgeman, J and Ferguson-Smith, Professor M (2000) *The BSE Inquiry Report. Evidence and Supporting Papers of the Inquiry into the Emergence and Identification of Bovine Spongiform Encephalopathy (BSE) and Variant Creutzfeldt-Jakob Disease (vCJD) and the Action Taken in Response to it up to 20 March 1996*, The Stationery Office, London, October, Vol 1, p1 – referenced simply as '*BSE Inquiry*'.
2   Cited by Narang, H (1997) *Death on the Menu*, H H Publishers, Newcastle.
3   *Farming Today* (2002) 'Interview with John Wilesmith', Radio 4, Birmingham, 12 September.
4   Lacey, Professor R (1998) *Poison on a Plate – the Dangers in the Food we Eat – and How to Avoid Them*, Metro, London, pp108, 227.
5   Van Zwanenberg, P and Millstone, E (2001) '"Mad Cow Disease" 1980s-2000: How Reassurances Undermined Precaution', in *Late Lessons from Early Warnings: the Precautionary Principle 1896–2000*, European Environment Agency, Copenhagen, p161.
6   *BSE Inquiry*, Vol 1, pp33–159;
7   Lacey (1998) op cit, p77.
8   Scott, P (2002) Speech to the Over Thirty Months (OTM) Rule Review, Public Meeting Organized by the Food Standards Agency, Congress Centre, London, 2 July; Hornsby, M (1996) 'EU Postpones Decision on Beef Byproducts', *The Times*, London, 11 April.
9   http://www.foodstandards.gov.uk/news/newsarchive/72117

10   Walker, J (2002) Speech to the Over Thirty Months (OTM) Rule Review, Public Meeting Organized by the Food Standards Agency, Congress Centre, London, London, 2 July.

11   *BSE Inquiry*, Vol 1, p164.

12   Smith, Professor P (2002) Speech to the Over Thirty Months (OTM) Rule Review, Public Meeting Organized by the Food Standards Agency, London, 2 July.

13   *BSE Inquiry*, Vol 1 pp157, 163–164; Godfrey, J (2002) Speech to the Over Thirty Months (OTM) Rule Review, Public Meeting Organized by the Food Standards Agency, London, 2 July.

14   Lacey, R (1998) 'We Have all Been Exposed to BSE Since 1986', in *Human BSE – Anatomy of a Health Disaster*, International Worker Books, Rotterdam, p42.

15   For details on BSE also see, http://www.nfu.org.uk/info/bse.asp#11

16   Walker (2002) op cit.

17   Firkins, L (2002) Speech to the Over Thirty Months (OTM) Rule Review, Public Meeting Organized by the Food Standards Agency, London, 2 July.

18   Walker (2002) op cit.

19   Godfrey (2002) op cit.

20   Meikle, J (2003) 'BSE Review Chief Hints that Ban on Older Cattle May End', *The Guardian*, 20 February, p14.

21   Dickinson, A (2001) *Response to Questions from the EFRA Select Committee*, 4 November.

22   Smith (2002) op cit.

23   Meikle, J (2002) 'Size of BSE Epidemic Worse than Thought', *The Guardian*, London, 9 October, p9.

24   Food Standards Agency (2002) *BSE Statistics*, Briefing Paper Produced for the OTM Meeting, 2 July.

25   Smith (2002) op cit.

26   http://www.bse.org.uk/files/yb/1988/07/06008001.pdf

27   www.foodstandards.gov.uk/multimedia/presentations/peter_smith.ppt

28   http://www.irlgov.ie/daff/AreasofI/BSE/BSE.htm

29   Mitchell, P (2001) 'World Health Organisation says BSE is a Major Threat', *World Socialist Web Site*, 6 July.

30   http://www.food.gov.uk/news/pressreleases/89130.

31   Smith (2002) op cit.

32   Ibid.

33   Boseley, S (2002) 'Tonsil Tissue Used for CJD', *The Guardian*, London, 20 September, p9.

34   Firkins, L (2003) *Chairman's Letter*, Human BSE Foundation, February.

35   Smith (2002) op cit.

36   Klarreich, E (2001) 'BSE's Epidemic Proportions', *Nature*, 23 November.

37   McKie, R et al (1996) 'A Conspiracy to Drive Us All Mad', *The Observer*, London, 24 March, pp16–17.

38   Godfrey (2002) op cit; Johnson, A (2001) *Variant CJD & Tonsillectomy – Where are we now?*, British Association of Otorhinolaryngologists, Head and Neck Surgeons, 29 August.

39   Klarreich, E (2001) 'BSE's Epidemic Proportions', *Nature*, 23 November.
40   http://www.bsereview.org.uk/data/sheep.htm
41   Meikle, J (2001) 'Study of BSE in Sheep Collapses in Tests Blunder', *The Guardian*, London, 19 October.
42   *BSE Inquiry*, Vol 1, pxix.
43   Bowcott, O et al (1996) 'Ministers Warned on Cattle Feed in 1980', *The Guardian*, London, 23 March, p5.
44   *BSE Inquiry*, Vol 1, pxix.
45   Dealler, S and Lacey, R (1990) 'Transmissible Spongiform Encephalopathies: the Threat of BSE', *Food Microbiology*, 7, p253–279.
46   Smith (2002) op cit.
47   Ibid.
48   *BSE Inquiry*, Vol 1, pxvii.
49   Dealler, S (2001) 'At Long Last, Signs of a BSE Breakthrough', *The Guardian*, London, 5 December, p16.
50   *BBC News Online* (1998) 'One in Three BSE Cases "Could Have Been Prevented",' 9 February; Highfield, R (1998) 'Vets "Wasted Two Years" in Battle Against BSE', *The Daily Telegraph*, London, 9 February.
51   *BSE Inquiry*, Vol 3, p4–5.
52   Harrison, M (1996) 'Three Treated for Suspected CJD', *PA News*, 25 April; quoted in Rampton, S and Stauber, J (1997) *Mad Cow U.S.A – Could the Nightmare Happen Here?*, Common Courage Press, Maine, pp91–92.
53   *BBC News Online* (1998), op cit.
54   Highfield, R (1997) 'The Woman who Discovered BSE', *The Daily Telegraph*, London, 9 February.
55   *BSE Inquiry*, Vol 3, p6.
56   *BBC News Online* (1998) op cit; *Associated Press* (1998) 'Disease Spread Due To Delay', 9 February.
57   *BSE Inquiry*, Vol 3, p8.
58   *BSE Inquiry*, Vol 1, p248.
59   Mead, N (2000) 'BSE Inquiry is Latest in Lord Phillips's High-Profile Career', *PA News*, London, 25 October.
60   *BSE Inquiry*, Vol 1, pxviii.
61   Dickinson, A (2001) Letter to the Chairman of the Agriculture Select Committee, House of Commons, 25 January.
62   *BSE Inquiry*, Vol 1, ppxvii–xviii.
63   *BSE Inquiry*, Vol 1, ExecSum.
64   Cordon, G (2000) 'BSE Report no "Whitewash" – Phillips', *PA News*, London, 26 October.
65   Carr-Brown, J et al (2000) 'A Culture of Secrecy that Risked our Lives', *The Sunday Times*, London, 29 October.
66   *Consumer Policy Review* (1006) Vol 6, No 3, May/June; quoted in Sheppard, J (1997) *From BSE to Genetically Modified Organisms – Science, Uncertainty and the Precautionary Principle*, Greenpeace, London, p24.
67   Lacey (1998) op cit, p77.
68   McKie et al (1996) op cit.
69   *The Financial Times* (1996) 27 March; quoted in Sheppard, J (1997) op cit.

70 McGill, I S (1998) *Statement to the BSE Inquiry.*
71 *BSE Inquiry*, Vol 1, pp34–35.
72 Connor, S (2001) 'Crucial Question that Holds Key to Future Safety', *The Independent*, London, 19 April, p7.
73 *BSE Inquiry*, Vol 3, pp122, 137–138.
74 Leather, S (1998) *Statement to the BSE Inquiry*, Number 248, date unknown.
75 Pennington, H (2000) 'The English Disease', *London Review of Books*, Vol 22, No 24, 14 December.
76 Rhodes, R (1998) *Deadly Feasts – Science and the Hunt for Answers in the CJD Crisis*, Touchstone, London, p179.
77 Rimmer, B (1998) *Statement to the BSE Inquiry*, Number 208, 21 September.
78 Hall, F (1998) *Statement to the BSE Inquiry*, Number 204, 31 March.
79 Rimmer (1998) op cit.
80 Toolis, K (2001) 'Epidemic in Waiting', *The Guardian*, Weekend, London, 22 September, p27.
81 Hornsby, M (1996) 'Mad Cow Disease an "Act of God"', *The Times*, London, 9 October.
82 Dickinson, A (1998) *Oral Evidence to the BSE Inquiry*, 11 June, p126.
83 *BSE Inquiry*, Vol 1, p100.
84 Smith, Sir J (1998) *Statement to the BSE Inquiry*, Number 181, p5.
85 Walford, D (1998) *Statement to the BSE Inquiry*, Number 182, p4.
86 Patterson, W (1998) *Statement to the BSE Inquiry*, Number 183, pp3–4.
87 *BSE Inquiry*, Vol 1, p233.
88 *BSE Inquiry*, Vol 16, pp5–8.
89 *BSE Inquiry* Vol 3, pp126–127.
90 Ibid, p128.
91 Ibid, p130.
92 *BSE Inquiry,* Vol 1, pp14, 47.
93 Lacey (1998) op cit, p103.
94 Dealler and Lacey (1990) op cit, pp253–279.
95 *BSE Inquiry*, Vol 16, pp5–8; Vol 3, p118; Vol 4, p2.
96 Lobstein et al (2001) *The Lessons of Phillips*, Thames Valley University; February.
97 Carr-Brown et al (2000) op cit.
98 Pennington (2000) op cit.
99 *BSE Inquiry*, Vol 4, p36.
100 Ibid, p68.
101 Ibid, p36.
102 Ibid, p62.
103 Lacey (1998) op cit, p111.
104 *BSE Inquiry*, Vol 4, p32.
105 Lacey (1998) op cit, p204.
106 *BSE Inquiry*, Vol 4, pp47, 68.
107 Van Zwanenberg and Millstone (2001) op cit, p160.
108 *BSE Inquiry*, Vol 1, p21.
109 http://www.bse.org.uk/files/yb/1989/06/02006001.pdf
110 Lacey (1998) op cit, p228.

111   Van Zwanenberg and Millstone (2001) op cit, p160.
112   *BSE Inquiry*, Vol 1, pxxii; www.bse.org.uk
113   *BSE Inquiry*, Vol 1, p117; www.bse.org.uk
114   Food Standards Agency (2002) *Agency Study Reports on Historic Use of Mechanically Recovered Meat in Food 1980–1995*, London, 10 October; Firkins, L (2002) *Chairman's Letter*, October.
115   BSE Inquiry, Vol 1, pp90, 118; www.bse.org.uk/files/yb/1990/08/00007001.pdf
116   *Farming Today* (2002) 'Interview with John Wilesmith', Radio 4, Birmingham, 12 September.
117   Smith (1998) op cit.

Chapter 3 Treated with Derision

1   Phillips, Lord, Bridgeman, J and Ferguson-Smith, Professor M (2000) *The BSE Inquiry Report. Evidence and Supporting Papers of the Inquiry into the Emergence and Identification of Bovine Spongiform Encephalopathy (BSE) and Variant Creutzfeldt-Jakob Disease (vCJD) and the Action Taken in Response to it up to 20 March 1996*, The Stationery Office, London, October, Vol 1, p234 – referenced simply as '*BSE Inquiry*', p1
2   *BSE Inquiry*, Vol 1, p52.
3   Dickinson, A (undated) *Statement to the BSE Inquiry*, Statement Number 74, p15.
4   *BSE Inquiry*, Vol 1, pp222–223.
5   Dickinson (undated) op cit, p15.
6   Dickinson, A (1999) *Comments on BSE Epidemiology Session and Some Associated Documents*, Evidence Submitted to the BSE Inquiry, 6 December.
7   Ibid.
8   Dickinson, A (undated) op cit, p3.
9   Dickinson, A (2001) *Letter to the Chairman of the Agriculture Select Committee*, House of Commons, Regarding Government Research into Transmissible Spongiform Encephalopathies and Intensive Farming, 25 January.
10   Dickinson, A (2001) *Response to Questions from EFRA Select Committee*, 4 November, p1.
11   Dickinson, (2001) op cit.
12   Ibid; Reid, H, Aitken, I, Dickinson, A, and Martin, W (2000) *Response to Scottish Executive Regarding the National Scrapie Eradication Plan*, October.
13   *BSE Inquiry*, Vol 3, pp147–148.
14   Holt, T (1998a) *Oral Evidence to the BSE Inquiry*, 31 March, p59.
15   Holt, T (1998b) *Statement for the BSE Committee of Inquiry*, 23 February.
16   Holt (1998a) op cit, pp88–90.
17   Fraser, H (1988) *Letter to Dr Tim Holt*, 9 December 1988.
18   Holt (1998a) op cit, p92.
19   Holt (1998b) op cit, 23 February.
20   Holt (1998a) op cit, p93.

21   Morgan, K (1999) *Statement to the BSE Inquiry*, Issued 22 September, Statement No 518.
22   Ibid.
23   Holt (1998a) op cit, p119.
24   Grant, H (1999) *Statement of Dr Helen Grant MD FRCP to the BSE Inquiry*, Statement Number 410, Issued 13 May.
25   *BSE Inquiry*, Vol 1, p110.
26   Grant (1999) op cit.
27   Martin, P (1995) 'The Mad Cow Deceit', *The Mail on Sunday*, London, Night and Day Section, 17 December.
28   Grant (1999) op cit.
29   Lacey, R (1998) *Poison on a Plate – the Dangers in the Food we Eat – and How to Avoid Them*, Metro, London, p100.
30   Ibid, p60.
31   Ibid, pp74, 117, 120, 125, 127–129.
32   *BSE Inquiry*, Vol 11, pp253–261.
33   Lacey (1998) op cit, pp151–153.
34   Ibid, pp131–139.
35   House of Commons Agriculture Committee (1990) *Bovine Spongiform Encephalopathy*, Fifth Report, 10 July.
36   Lacey (1998) op cit, pp131–137, 207; *BSE Inquiry* Vol 11, p243.
37   Lacey (1998) op cit, p208.
38   *BSE Inquiry*, Vol 11, p276.
39   Dealler, S (1998) *Statement to the BSE Inquiry*, Number 22, 26 March, p2.
40   Ibid.
41   *BSE Inquiry*, Vol 11, pp245–247, 275.
42   Ibid, p256.
43   Ibid, p268.
44   Ibid, p266.
45   Dealler, S (2001) 'At Long Last, Signs of a BSE Breakthrough', *The Guardian*, London, 5 December, p16.
46   Gajdusek, C (1985) *Letter to Dr J Whitehead*, Director of the Public Health Service Laboratory, National Institutes of Health, Maryland, 15 May; BSE Inquiry Number M37/85/5.15/1.1–1.2.
47   Narang, H (2002) Interview with Author, August; Narang, H (1998a) 'The Government has Deliberately Sabotaged Science', in *Human BSE – Anatomy of a Health Disaster*, International Worker Books, Rotterdam, p130.
48   Narang, H (1998b) *Statement to the BSE Inquiry*, Number 113, p14.
49   Ibid, p15.
50   Public Health Laboratory Service Board (1998) *Comments on a Submission by Dr Harash K Narang*, Statement No 114, pp3–4.
51   Narang, H (1998b) op cit, p19; Public Health Laboratory Service Board (1998), p5.
52   McKie, R et al (1996) 'A Conspiracy to Drive Us All Mad', *The Observer*, London, 24 March, pp16–17.
53   Narang, H (1998a) op cit, p131.
54   Ibid, p128; Martin (1995) op cit.

55   Bailey, J. A (1995) *Collaboration Agreement with Electrophoretics International PLC*, memo to F. Strang, 17 March: BSE Inquiry Document 95/3.17/2.2.
56   *BSE Inquiry*, Vol 1, p280.
57   Clark, D (1999) *Statement to the BSE Inquiry*, Number 428, 2 June, p4.
58   Narang, H (1998b) op cit, pp17–20.
59   Public Health Laboratory Service Board, op cit, p11.
60   Narang, H (1997) *Death on the Menu*, H H Publishers, Newcastle, p79.
61   Narang (1998b) op cit, p9.
62   Narang. H (2001) 'A Critical Review of Atypical Cerebellum-Type Creutzfeldt-Jakob Disease: Its Relationship to 'New Variant' CJD and Bovine Spongiform', *Journal of the Society for Experimental Biology and Medicine*, Vol 226, No 7, pp629–639.
63   BBC Radio Four News (2002) The Today Programme, 16 August.
64   Narang, H (1998b) op cit, p55.
65   Narang, H (1997) *The Link – From Sheep to Cow to Man*, H H Publisher, Newcastle, p179.
66   *BSE Inquiry*, Vol 1, pxix.
67   Smith, P (2002) OTM Meeting, London, 2 July.
68   Sibley, D (2002) OTM Meeting, London, 2 July.
69   Monbiot, G (2000) 'Mad Cows, Bretons and Manganese', *The Guardian*, London, 23 November.
70   Whatley, S (1998) *Statement to the BSE Inquiry*, Statement Number 21, pp2–3.
71   Forde Gracey, J (1998) *Statement to the BSE Inquiry*, Statement Number 130, p 13.
72   The Working group was called BSEP or the Biology of the Spongiform Encephalopathies Programme – BBSRC (1995) Coordinated Programme of Research on the Biology of the Spongiform Encephalopathies – In Confidence, Meeting of the BSEP Working Party, 27 January; BSE Inquiry Document Number 95/1.27/1.1.
73   Horn et al (2001) *Review of the Origin of BSE*, MAFF, 2001, 5 July, p37.
74   Ibid, p36.
75   Ibid, p36; Purdey, M (2001) *The Prof. Gabriel Horne Report on the Origins of BSE and The Research of Mark Purdey*, December.
76   Boseley, S (1996) 'How the Truth Was Buried', *The Guardian*, London, 24 March, p23.

Chapter 4 Silent Spread

1   Beat, A and Beat R (2002a) *Submission to the Lessons Learned Inquiry*, Bridge Mill, Devon, 11 March.
2   Brown, F (1998) *Oral Evidence to the BSE Inquiry*, 24 March.
3   Barlow, R (1999) *Statement to the BSE Inquiry*, Statement Number 565, 7 December, p7.
4   Anderson, I (2002) *Foot and Mouth 2001: Lessons to be Learned Inquiry*, p41; Beat, A (2001) *Foot and Mouth – The Facts*, on www.smallholders.org/foot and mouth/ foot and mouth %20An%20Overview.rtf

5   Department of Health (2001) *A Rapid Qualitative Assessment of Possible Risks to Public Health from Current Foot & Mouth Disposal Options*, Main Report, London, June, p6.
6   http://www.tetracore.com/profile.html
7   Anderson (2002) op cit, p50; Scudamore, J (2002) *Origin of the UK Foot and Mouth Disease Epidemic in 2001*, Defra, London, June, p5.
8   Waugh was found guilty of failing to notify the authorities of foot and mouth disease and inflicting unnecessary cruelty on his animals.
9   Burke, J et al (2001) 'Virus Alert ignored for Three Years', *The Observer*, London, 25 March, p1.
10   Anderson (2002) op cit, pp8, 42.
11   Leake, J (2001) 'Foot and Mouth: The Blunders', *The Sunday Times*, London, 23 December, p7.
12   Donaldson, A (2002) *Interview with Bonnie Durrance and Alan Beat*, Pirbright, 3 April.
13   The Comptroller and Auditor General (2002) *The 2001 Outbreak of Foot and Mouth*, The National Audit Office, June, p1; *The Guardian* (2002) 'Back on the Menu: First British Beef Exports Since Ban is Lifted', London, 26 September, p1.
14   The Comptroller and Auditor General (2002) op cit, p1.
15   Figures from Jane Connor at the Meat and Livestock Commission, circulated on Smallholders Online Newsletter (2002) Number 3, 4 May.
16   Beat, A and Beat, R (2002b) Interview with Author, September.
17   Booker, C and North, R (2001) 'Not the Foot and Mouth Report', *Private Eye*, London, November, p30.
18   *Western Morning News* (2001) 'Foot and Mouth – How The West Country Lived Through the Nightmare', Plymouth, p3.
19   Leake (2001) op cit.
20   *Western Morning News* (2001) op cit. p78.
21   Leake (2001) op cit.
22   *Western Morning News* (2001) op cit, pp9–10, 70
23   Anderson (2002) op cit, p26.
24   *Western Morning News* (2001) op cit, p3.
25   *Western Morning News* (2001) Editorial, 22 May.
26   Brown, F (2002) Interview with Author, September; Callahan, J et al (2002) 'Use of a Portable Real-Time Reverse Transcriptase-Polymerase Chain Reaction Assay for Rapid Detection of Foot-and-Mouth Disease Virus', *Journal of the American Veterinary Medical Association*, 1 June, Vol 220, No 11, pp1636–1642.
27   Intervet (2001) *Intervet Announces New Test for Foot and Mouth*, Press Release, Boxmeer, The Netherlands, 3 April.
28   Intervet (2001) *Intervet Announces Foot and Mouth Marker Test*, Press Release, Boxmeer, The Netherlands, 11 October.
29   King, D (2001) *Evidence to the Select Committee on Agriculture*, London, 25 April.
30   Donaldson, A (2001) *Evidence to the Select Committee on Agriculture*, 25 April.
31   Booker and North (2001) op cit, p9.

230 Don't Worry – It's Safe to Eat

32   National Federation of Badger Groups (undated) *Summary of Risks to Human Health from Bovine Tuberculosis*; www.nfbg.org.uk/

33   Booker and North (2001) op cit, p9; Lodger, N (2000) 'Top UK Epidemiologist Suspended after Complaints', *Nature*, 27 January, Vol 403, p353; Lodger, N (2000) 'Anderson Steps Down From Wellcome over Oxford Row', *Nature*, 3 February, Vol 403, p472; Lodger, N (2000) 'Oxford Professor Faces Business Link Inquiry', *Nature*, 17 February, Vol 403, p695.

34   Foot and Mouth Disease Lessons Learned Inquiry (2002) *Note of Meeting*, Present: Professor Roy Anderson – Imperial College, Dr Iain Anderson, Inquiry Chairman; Alun Evans, Secretary to the Inquiry, London, 11 June.

35   Booker and North (2001) op cit, p9.

36   Ibid; Beat and Beat (2002b), *Interview with Author*, September.

37   Government Information and Communications Service (2001) *Transcript of BBC Newsnight*, Media Monitoring Unit, 21 March; Booker and North (2001) op cit, p7.

38   *Private Eye* (2001) 'Down on the Farm,' London, No 1029, 1–14 June, p12.

39   Beat and Beat (2002b) *Interview with Author*, September.

40   Booker and North (2001) op cit, p11.

41   Donaldson (2002) op cit.

42   Royal Society (1999) *Letter From a Group of Fellows at The Royal Society*, London, 22 February.

43   Kitching, R P (2002) *Submission to the Temporary Committee on Foot-and-Mouth Disease*, 16 July.

44   Booker and North (2001) op cit, p24.

45   Ferguson, N, Donnelly, C, and Anderson, R (2001) 'The Foot-and-Mouth Epidemic in Great Britain: Pattern of Spread and Impact of Interventions', *Science*, Vol 292, 11 May, pp1157–1158.

46   Hutber, A M and Kitching, R P (1996) 'The Use of Vector Transmission in the Modelling of Intraherd FMD', *Environmental and Ecological Statistics* Vol 3, pp245–255; Hutber, A M and Kitching, R P (1998) 'Control of FMD Through Vaccination and the Isolation of Infected Animals', *Tropical Animal Health and Production*, Vol 3, pp217–227; Hutber, A M, Kitching, R P and Conway, D A (1999) 'Predicting the Level of Herd Infection for Outbreaks of FMD in Vaccinated Herds', *Epidemiology and Infection*, Vol 122, pp539–544; Hutber, A M and Kitching, R P (2000) 'The Role of Management Segregations in Controlling Intra-herd FMD', *Tropical Animal Health and Production*, Vol 32, pp285–294.

47   Beat and Beat (2002b) *Interview with Author*, September.

48   Ferguson, Donnelly and Anderson (2001) op cit, p1156.

49   Ferguson, N, Donnelly, C and Anderson, R (2001) 'Transmission Intensity and Impact of Control Policies on the Foot and Mouth Epidemic in Great Britain', *Nature*, Vol 413, 4 October, pp544, 547.

50   Beat, A (undated) Modelling Reexamined, Devon.

51   Williams, B and Kuhn, A (2001) Why Did Healthy Cattle Have to Die? *Western Morning News*, Plymouth, 24 May, pp22–23.

52   *Western Morning News* (2001) 'Animal Health Bill Must be Opposed', Editorial, 12 December, Plymouth, p10.

53   Donaldson (2002) op cit.

54   *Western Morning News,* op cit, p69.

55   Leppard, D and Ungoed-Thomas, J (2001) 'Top Scientist Condemns "Flawed" Cull', *The Sunday Times,* London, 29 April.

56   Kitching, R P (2002) *Submission to the Temporary Committee on Foot-and-Mouth Disease,* 16 July.

57   Ibid.

58   Trewin, C (2001) 'Brown Defiant Over Culling', *Western Morning News,* Plymouth, p20.

59   Linklater, M (2001) 'Wrong, Wrong, Wrong, Wrong', *The Times,* London, Section 2, 24 May, pp2–3.

60   Kitching (2002) op cit.

61   Adams, T (2002) 'Death in the Countryside', *The Observer,* Life Magazine, London, 10 February, p22.

62   Kitching (2002) op cit.

63   Osborn, A (2002) 'MEPs Slate Foot and Mouth Cull', *The Guardian,* London, p4.

64   Vidal, J (2001) 'Green Allies Demand Vaccination Option', *The Guardian,* London, 12 April, p6.

65   http://www.guardian.co.uk/footandmouth/story/0,7369,464865,00.html

66   Booker, C (2001) 'Now a U-Turn Over Vaccination?', *The Daily Mail,* London, 28 March, p6.

67   Heart of Cumbria Campaign (2001) *Public Meeting to Discuss Vaccination Against Foot and Mouth Disease,* Penrith, 16 September.

68   Ahmed, K et al (2001) 'Operation Cobra: Can it Stop Disaster'? *The Observer,* London, 25 March, pp9–11.

69   North, R (2001) *The Death of British Agriculture,* Duckworth, London, p6.

70   Beat and Beat (2002b) *Interview with Author,* September; Booker and North (2001) op cit, p14.

71   Anderson (2002) op cit, p7.

72   Booker and North (2001) op cit, p28.

73   Wintour, P and Chrisafis, A (2001) 'Countryside Lobby Attacks Inquiry "Secrecy"', *The Guardian,* London, 10 August, p7.

74   Ibid.

75   Trewin, C (2001) 'Animal Health Bill Makes "Flawed Assumptions"', *Western Morning News,* West Country Farming Section, Plymouth, 12 December, p1.

76   Defra (2002) *FMD Inquiries; Defra Response,* Press Release, London, 6 November.

77   The Comptroller and Auditor General (2002) op cit, p94.

78   McDougall, D (2002) 'A Dirty Business?' *The Scotsman,* 21 August.

79   Allardyce, J (2002) 'Scots Firm in Labour Donation Probe', *The Scotsman,* 24 February.

80   Reid, G (2000) *Interview with Author,* autumn; Transport and the Environment Committee (2002) *Extract From The Minutes,* Scottish Parliament, Edinburgh, 16 January; Annex C: Written Evidence; Petition PE327 By The Blairingone And Saline Action Group; Hope, D (2002) *Interview with Author,* September.

81   Transport and the Environment Committee (2002) op cit.
82   Scottish Environmental Protection Agency (2001) *Snowie Ltd Fined £5000 For Disposing of Waste in a Manner Likely to Cause Pollution of the Environment*, Press Release, Stirling, 22 June; Scottish Environmental Protection Agency (2000) *Snowie Limited Fined £3000 for Pollution Incident*, Press Release, Stirling, 23 February; Transport and the Environment Committee (2002) *Report on Petition PE327 by the Blairingone and Saline Action Group On Organic Waste Spread On Land*, 4th Report, Scottish Parliament, Edinburgh.
83   Edwards, R (2000) 'Animal Blood and Guts Dumped on Farms, *The Sunday Herald*, Glasgow, 19 November; *BBC News Scotland* (2002) 'Blood on the Land' Ban Demand, 26 June.
84   Scottish Environment Protection Agency (1998) *Strategic Review of Organic Waste on Land*, Stirling, October, p4.
85   Hope, D (2002) *Scottish Environment Protection Agency – Policy and Financial Management Review*, Submission from the Blairingone and Saline Action Group, September, p7.
86   Scottish Environment Protection Agency (1998) op cit, p11.
87   Snowie (undated) *Material Driven onto Lambhill Site Since August 1999 up to August 2000*.
88   Transport and the Environment Committee (2002) op cit.
89   Transport and the Environment Committee (2001) *Official Report*, Scottish Parliament, Edinburgh, 3 October.
90   Scottish Parliament (2002) *Parliamentary Committee Backs Blairingone Petition For New Waste-Spreading Regulations*, Committee News Release, Edinburgh, 7 March.
91   Transport and the Environment Committee (2001) *Agenda*, 3 October; Submissions on Petition PE327 by the Blairingone and Saline Action Group.
92   Transport and the Environment Committee (2001) op cit.
93   Matthews, D (1989) *Letter to Derek Wilson*, MAFF, Surbiton, 5 June; BSE Inquiry Document Number 89/06.05/9.1-9.2.
94   Spongiform Encephalopathy Advisory Committee (2002) *Minutes of the 72nd Meeting*, Defra, London, 6 February.

Chapter 5 Hot Potato

1   10 Downing Street (1997) Press Notice regarding the Proposals of Setting up the FSA, 8 May.
2   Pusztai, A (2000) 'Dr Pusztai Speaks', *Laboratory News*, February.
3   Pusztai, A (2002) 'GM Food Safety: Scientific and Institutional Issues', *Science as Culture*, Number 1, p69–92.
4   Interview with Michael Gillard, 1999.
5   Pusztai, A (2001) *Lecture given at Schumacher College*, Devon, January.
6   Flynn, L, Gillard, M and Rowell, A (1999) 'Ousted Scientists and the Damning Research into Food Safety', *The Guardian*, London, 12 February, p6.

7   Pusztai (2000) op cit.

8   Pusztai, A (2002) *The 'Scientific Advisory System: Genetically Modified Foods' Inquiry*; Memorandum to the Science and Technology Committee, 1999, p1.

9   Science and Technology Committee (1999) *Scientific Advice to Government: Genetically Modified Food*, Memorandum Submitted by the Rowett Research Institute, The Stationery Office, London, 8 March, p41.

10   Flynn, Gillard and Rowell (1999) op cit.

11   Pusztai, A (undated) *Bt* Maize Story, Notes; Flynn, Gillard and Rowell (1999) op cit.

12   The Rowett Research Institute, *Annual Report 1994*, Aberdeen, 1995, pp126–127.

13   Professor James, P (2002) *Interview with Author*, September.

14   House of Commons Select Committee on Agriculture (1998) *Fourth Report*, 22 April; quoting Food Standards Agency: an interim proposal by Professor Philip James, 30 April 1997.

15   James, P (1998) Interviewed on *Forbidden Fruit*, Frontline, BBC Scotland, February.

16   World in Action (1998) *New Health Fears Over 'Frankenstein' Food*, Press Release, Granada Television Limited, 10 August.

17   Verakis, D (1998) Speaking on GMTV, 10 August, 6.25 a.m.

18   Rowett Research Institute (1998) *Genetically Modified Foods*, Press Release, Aberdeen, 10 August.

19   Rowett Research Institute (1998) *Genetically Modified Foods*, Press Release, Aberdeen, 10 August.

20   Science and Technology Committee (1999) *Scientific Advice to Government: Genetically Modified Food, Minutes of Evidence Dr Arpad Pusztai and Dr Stanley Ewen and the Rowett Research Institute*, Professor Philip James and Dr Andrew Chesson, The Stationery Office, London, 8 March, p47.

21   Verakis(1998) op cit.

22   Pusztai, A (1998) Speaking on GMTV, 10 August, 6.25 am.

23   Ingham, I (1998) 'Genetic Crops Stunt Growth', *The Express*, London, 10 August, p1.

24   Cunningham, J (1999) Interview on TV News and Channel Four News, 12 February.

25   Professor Burke, D (1999) 'No Big Deal', *The Guardian*, London, 13 February, p21.

26   Pusztai, A (1998), Speaking on World in Action, *How Safe is Genetically Modified Food*, 10 August.

27   Rowett Research Institute (1998) *Genetic Manipulation of Foods*, Press Release, Aberdeen, 11 August.

28   Bardocz, S (1999) Summary of Research Results, 5 Page Note to Professor James, Professor Bremner and Dr Chesson, 11 August.

29   BBC (1998) 'UK Genetics Scientist Suspended', report by Science Correspondent James Wilkinson, 12 August.

30   Hawkes, N, (1998) 'Scientist's Potato Alert Was False, Laboratory Admits', *The Times*, London, 13 August.

31 Wilson, E (1998) 'Got It Wrong', *The Daily Mail*, London, 13 August.
32 Radford, T (1998) 'Scientist in Genetic Food Scare Suspended', *The Guardian*, London, 13 August.
33 Science and Technology Committee (1999) op cit, p47
34 Kirby, A (1999) 'GM Foods: What's the Hurry', BBC, London, 17 February.
35 *BNA Chemical Regulation Daily* (1995) 'Administration Committed to Industry R&D Despite Budget Cuts, Gore Adviser Says', 13 January.
36 Lambrecht, B (2001) 'Dan Glickman – Outgoing Secretary Says Agency's Top Issue is Genetically Modified Food', *St. Louis Post-Dispatch*, 25 January.
37 St. Clair, J (1999) 'The Monsanto Machine', *In These Times*, 7 March.
38 Ibid.
39 Eastham, P (1999) 'They Couldn't be Closer to Blair. So Why are These Men Working for the World's Biggest Genetic Food Firm and Opening Doors to the Highest Level of Government?' *The Daily Mail*, London, 13 February, p5.
40 The Royal Society of Canada (2001) Expert Panel on the Future of Food Biotechnology, Ottawa, p179.
41 *BBSRC Business* (1998) 'Making Crops Make More Starch', January, p6–7; quoted in Diamond, E (2001) *The Great Food Gamble – An Assessment of Genetically Modified Food Safety*, Friends of the Earth, London, May, p14.
42 Flynn, Gillard and Rowell (1999) op cit.
43 Martineau, B (2000) *First Fruit – The Creation of the Flavr Savr™ Tomato and the Birth of Biotech Food*, McGraw Hill, New York, p69.
44 Dr Pozueta-Romero, J (2002) 'Tools of Genetic Engineering in Plants', in *Fruit and Vegetable Biotechnology*, ed Victoriano Valpuesta, Woodhead Publishing, Cambridge, March, Section 2.6.1, p12.
45 Senior, I and Dale, P (1996) 'Plant Transgene Silencing – Gremlin or Gift?' *Chemistry & Industry*, 19 August, p605.
46 Pozueta-Romero (2002) op cit, p10.
47 Finnegan, H and McElroy D (1994) 'Transgene Inactivation: Plants Fight Back!,' *Bio/Technology*, 12, p883.
48 Ibid.
49 Pozueta-Romero (2002) op cit, p10.
50 Gillard, M S and Flynn, L (1999) 'Key GM Gene is Owned by Monsanto', *The Guardian*, 17 February, pp1.
51 Ibid.
52 See the Genewatch UK database at www.genewatch.org
53 Hansen, M (2000) *Genetic Engineering is Not an Extension of Conventional Plant Breeding; How Genetic Engineering Differs from Conventional Breeding, Hybridization, Wide Crosses and Horizontal Gene Transfer*, Consumer Policy Institute/Consumers Union, January.
54 Al-Kaff, N et al (1998) 'Transcriptional and Posttranscriptional Plant Gene Silencing in Response to a Pathogen', *Science*, Vol 279, 27 March, p2113; Pozueta-Romero (2002) op cit.
55 Hansen (2000) op cit.

56   Coghlan, A (2002) 'Don't Panic', *New Scientist*, Vol 174, No 2340, 27 April, p13.

57   Martin-Orue, S et al (2002) *British Journal of Nutrition*, in press at time it was quoted by Netherwood, T et al (2002) *Transgenes in Genetically Modified Soya Survive Passage Through the Human Small Bowel but are Completely Degraded in the Colon*, Department of Biological and Nutritional Sciences and Department of Agricultural and Environmental Sciences, University of Newcastle upon Tyne.

58   University of Newcastle (2002) *Technical Report on the Food Standards Agency Project G010008 'Evaluating the Risks Associated with Using GMOs in Human Foods'*, Food Standards Agency, July.

59   James, P (1999) Interviewed on Reporting Scotland, 12 February.

60   Rhodes, J (1999) Re: SOAEFD Flexible Fund Project RO 818 – Audit Data 21.08.98, Letter to Arpad Pusztai, 10 February.

61   Pusztai (2000) op cit.

62   Bourne, Chesson, A, Davies, H and Flint, H (1998) *SOAEFD Flexible Fund Project RO 818, Audit of Data Produced at the Rowett Research Institute*, 21 August.

63   Rhodes, J (1999) Letter to Arpad Pusztai, Regarding SOEFD Flexible Fund Project RO 818, 10 February.

64   Ibid.

65   Bourne et al (1998) op cit; Rowett Research Institute (1998) *Genetically Modified Organisms: Audit Report of Rowett Research on Lectins*, Press Release, Aberdeen, 28 October.

66   James, P (1998), *Evidence Before the House of Lords Select Committee*, London, 28 October.

67   Memorandum signed by 30 Scientists (1999) February. The scientists were Professor K Baintner, Department of Physiology, Pannon Agricultural University, Kaposvar, Hungary ; Profs. B S Cavada, R de Azevedo Moreira, A F F U de Carvalho, M de Guia Silva Lima, J T A de Oliveira, I M Vasconcelos (previous PhD students and/or collaborators of Pusztai) Universidade Federal do Ceara, Fortaleza, Brazil; Prof J Cummins, Emeritus Prof Genetics, Ontario, Canada; Dr S W B Ewen, Department of Pathology, Aberdeen Royal Hospitals, Aberdeen; Prof R Finn, Department of Medicine, The University of Liverpool; Prof M Fuller, Stony Brook, NY 11790; Prof B C Goodwin, Schumacher College, Dartington, Devon; Dr J Hoplichler, Federal Institute for Less-Favoured and Mountainous Areas, Vienna; Dr C V Howard, Fetal and Infant Toxico-Pathology, The University of Liverpool, UK; Dr J Koninkx, Department of Pathology, Faculty of Veterinary Medicine, University of Utrecht; Prof A Krogdahl, Norwegian School of Veterinary Science, Oslo; Dr K Lough, Bankhead, Aberdeen, (formerly of the Rowett Research Institute, Aberdeen; Dr D Mayer, Heidelberg; Prof F V Nekrep, Biotechnical Faculty, Zootechnical Department, University of Ljubljana, Slovenia; Prof S Pierzynowski, Department of Animal Physiology, University of Lund, Sweden; Prof S Pongor, Protein Structure and Function Group, International Centre for Genetic Engineering and Biotechnology, Trieste,

Italy; Prof I Pryme, Department of Biochemistry and Molecular Biology, University of Bergen, Norway; Prof J Rhodes, Gatroenterology Research Group, The University of Liverpool; Dr L Rubio, Department of Animal Nutrition, Estacion Experimental del Zaidin, Granada, Spain; Prof M Sajgo, Department of Chemistry and Biochemistry, Godollo University of Agriculture, Hungary; Prof U Schumacher, Department of Neuroanatomy, University of Hamburg, Germany; Dr B Tappeser, Institute for Applied Ecology, Freiburg, Germany; Prof T Wadström, Department of Medical Microbiology, University of Lund, Sweden; E Van Driessche, Prof, Laboratory of Protein Chemistry, Vrije Universiteit Brussel, Brussels, Belgium; T C Bøg-Hansen, senior associate professor, The Protein Laboratory, Institute of Molecular Pathology, University of Copenhagen, Denmark

68   Flynn, Gillard and Rowell (1999) op cit, p7.
69   Leake, C and Fraser, L (1999) 'Scientist in Frankenstein Food Alert is Proved Right', *The Mail on Sunday*, London, 31 January.
70   Rowett Research Institute (1999) 'Genetically Modified Foods', Press Release, Aberdeen, 12 February.
71   Birch A et al (1999) 'Tri-trophic Interactions Involving Pest Aphids, Predatory 2-Spot Ladybirds and Transgenic Potatoes Expressing Snowdrop Lectin for Aphid Resistance', *Molecular Biology*, Vol 5, pp75–83.
72   Down, R et al (2000) 'Snowdrop Lectin (GNA) Has No Acute Toxic Effects on a Beneficial Insect Predator, the 2-Spot Ladybird (*Adalia Bipunctata* L.)', *Journal of Insect Physiology*, Vol 46, pp379–391.
73   Birch A et al (2002) The Effect of Genetic Transformations for Pest resistance on Foliar Solanidine-Based Glycoalkaloids of Potato (*Solanum tuberosum*), *Annals of Applied Biology*, Vol 140, pp143–149.

Chapter 6 The 'Star Chamber'

1   Cabinet Office (1999) *GM Foods and Crops: Policy and Presentation Issues*, Restricted, 19 February.
2   Royal Society, (1999) *Royal Society Experts to Probe GM Food Concerns*, Press Release, London, 19 February
3   Klug, Sir A (1999) *Speech Given to the Annual Luncheon of the Parliamentary and Scientific Committee*, London, 23 February.
4   Royal Society (1999) *Letter From a Group of Fellows at The Royal Society*, London, 22 February.
5   Meek, J (20001) 'Scientific Elite Put Under Microscope', *The Guardian*, London, 17 March, p15.
6   House of Commons Science and Technology Committee (2002) *Government Funding Of the Scientific Learned Societies, Fifth Report of Session 2001-02, Volume 1: Report and Proceedings of the Committee*, 1 August, p44.
7   Leake, J (2002) 'Royal Society Found Guilty of Keeping out Woman Scientists', *The Sunday Times*, London, 28 July, p7.

8   House of Commons Science and Technology Committee (2002) op cit, pp40–43.

9   Wakeford, T (2000) 'Genetically-Modified Science', *Times Higher Educational Supplement*, April.

10   Wakeford, T (2002) *Personal Communication with Author*, October.

11   Collins, P (2002) *Communication with Author*, The Royal Society, London, 6 November.

12   See Bøg-Hansen's website http://plab.ku.dk/tcbh/Pusztaitcbh.htm; Pusztai, A (2002) *Interview with Author*, November.

13   Ibid.

14   Cox, S (1999) *Letter to Arpad Pusztai*, London, 15 March.

15   Pusztai, A (1999) *Letter to Dr Rebecca Bowden*, Aberdeen, 19 March

16   Pusztai, A (1999) *Letter to Rebecca Bowden*, Aberdeen, 19 March.

17   Ewen, S (1999) *Re Royal Society Involvement in the Debate of the Safety of GM Food*, Letter to Stephen Cox, Royal Society, 23 March.

18   Cox, S (1999) *Letter to Arpad Pusztai*, London, 23 March.

19   Anonymous (1999) Hand-written notes, 31 March – the fact that the reviewer had already commented was removed from the typed version Pusztai received in an email from Rebecca Bowden on 10 May 1999.

20   Anonymous (1999) *Review of Data on Toxicity and Allergenicity of GM Food*, Sent to Pusztai in an email from Rebecca Bowden on 4 May.

21   Anonymous (1999) GM Potatoes, 8 April.

22   Anonymous (1999) *Possible Toxicity of GM potatoes, Comments on the Work by Professor Arpad Pusztai*, 12 April.

23   Anonymous (1999) comments enclosed in an email sent by Rebecca Bowden to Professor Pusztai, 10 May.

24   Ibid.

25   Pusztai, A (1999) *Email to Rebecca Bowden*, Aberdeen, 4 May.

26   Bowden, R (1999) *Email to Professor Pusztai*, London, 10 May; Pusztai, A (2002) *Interview with Author*, March; Collins (2002), op cit.

27   Bowden, R (1999) *Email to Arpad Pusztai*, London, 13 May.

28   Pusztai, A (1999) *Letter to Stephen Cox*, Aberdeen, 12 May.

29   The Royal Society (1999) *Review of Data on Possible Toxicity of GM Potatoes*, London, 17 May.

30   Collins (2002) op cit.

31   Radford, T (1999) 'GM Research "Flawed in Design and Analysis"', *The Guardian*, London, 19 May, p11; Derbyshire, D (1999) 'Experts Say Key Research is Flawed', *The Daily Mail*, London, 19 May, p6.

32   Pusztai, A et al (1999) 'Expression of the Insecticidal Bean α-Amylase Inhibitor Transgene Has Minimal Detrimental Effect on the Nutritional Value of Peas Fed to Rats at 30% of the Diet', *Journal of Nutrition*, Vol 129, pp1597–1603.

33   Ibid.

34   Science and Technology Committee (1999) *Scientific Advisory System: Genetically Modified Foods*, The Stationery Office, London, First Report, Volume 1, pxv.

35   Ibid, p30.

36   Pusztai, A (2002) *Interview with Author*, March.
37   Pusztai, A (2002) *The 'Scientific Advisory System: Genetically Modified Foods' Inquiry*; Memorandum to the Science and Technology Committee, 1999, p2.
38   Horgan. G W and Glasbey, C A (1999) *Statistical Analysis of Experiments on Genetically Modified Potatoes at the Rowett Research Institute*, Biomathematics and Statistics, Scotland, 1 March.
39   Quotes taken from different news broadcasts on the 18 May 1999.
40   Office of Science and Technology (1999) *The Advisory and Regulatory Framework for Biotechnology: Report from the Government's Review*, Cabinet Office, May.
41   Britton, P (1999) *Biotechnology Presentation Group*, Letter to John Fuller, Private Secretary to the Cabinet Office, 11 May.
42   Cunningham, J (1999) Speaking in the House of Commons; 21 May.
43   Science and Technology Committee (1999) op cit, pxvi.
44   Collins (2000) op cit.
45   Booker, C and North, R (2001) 'Not the Foot and Mouth Report', *Private Eye*, London, p28.
46   Cunningham (1999) op cit.
47   *The Lancet* (1999) 'Health Risks of Genetically Modified Foods,' Editorial, London, Vol 353, No 9167, 29 May, p1811.
48   Cox, S (1999) *Letter to Alan Rusbridger*, London, 25 May.
49   Collins, P (1999), *Letter to Michael Gillard*, London, 24 May.
50   Cox, S (1999) *Letter to Michael Gillard*, London, 17 June.
51   Letter from a Group of Fellows at The Royal Society, 22 February 1999, signed by Brian Heap; Patrick Bateson; Sir Eric Ash; Roy Anderson; Sir Alan Cook; Sir Roger Elliott; Professor William Hill; Louise Johnson; Sir John Kingman; Peter Lachmann, Dr Paul Nurse; Linda Partridge; Dr Max Perutz; Sir Martin Rees; Sir Richard Southwood; Sir John Meurig Thomas; Sir Ghillean Prance; Lord Selbourne; Robert White.
52   Royal Society (1998) *Genetically Modified Organisms – the Debate Continues*, Press Release, London, September.
53   Nuffield Council on Bioethics (1999) *Genetically Modified Crops: the Ethical and Social Issues*, Nuffield Council, London, p136.
54   The Working Group was chaired by Professor Noreen Murray, Professor of Molecular Genetics at the Institute of Cell and Molecular Biology, University of Edinburgh, whose research is funded by the Medical Research Council. Murray's Department has links to, amongst others, the Roslin Institute and the Scottish Crop Research Unit. Also from the University of Edinburgh, was Professor William Hill, from the Institute of Cell, Animal and Population Biology at the Division of Biological Sciences. Much of Hill's work is 'in collaboration with the Roslin Institute and the animal breeding industry'. According to the Roslin's annual report, Hill was Deputy Chairman of the Roslin and on its Governing Council. Michael Waterfield was the Head of Department of Biochemistry and Molecular Biology, University College London. In 1999 scientists at his department were funded by Unilever, GlaxoWellcome, Glaxo and Glaxo

Group Research, amongst others (Cartermill International, *Current Research in Britain – Biological Sciences*, 13th Edition, pp216–218). Professor Brian Heap had been Director of Research at Roslin and another key biotechnology institute, Babraham. Heap had also held consultancies in the pharmaceutical sector.

55   Connor, S (1999) 'Scientists Revolt at Publication of "Flawed" GM Study', *The Independent*, London, 11 October, p5.

56   Highfield, R and Aisling, I (1999) 'Study is Published and Damned', *The Daily Telegraph*, London, 13 October, p 21.

57   BBSRC (1999) *BBSRC Concern About GM Paper in The Lancet*, BBSRC Press Release, Swindon, 13 October.

58   Flynn, L and Gillard, M S (1999) 'Pro-GM Food Scientist "Threatened Editor"', *The Guardian*, London, 1 November.

59   Lachmann, P (1999) 'Health Risks of Genetically Modified Foods', *The Lancet*, London, Vol 354, 3 July, p69.

60   Ibid.

61   Flynn and Gillard (1999) op cit.

62   Ibid.

63   Ibid.

64   Gillard, M S, Rowell, A, Flynn, M (1999), Draft of Copy, Legalled and Changed, 11.30 am, 26 October.

65   Ibid.

66   Horton, R (1999) *Interview with Author and Michael Gillard*, October.

67   Ewen, S W B and Pusztai, A (1999) 'Effects of Diets Containing Genetically Modified Potatoes Expressing *Galanthus Nivalis* Lectin on Rat Small Intestine', *The Lancet*, London, Vol 354, 16 October, pp1353–1354.

68   Horton, R (1999) 'Genetically Modified Foods: "Absurd" Concern or Welcome Dialogue,' *The Lancet*, London, Vol 354, 16 October, pp1314–1315.

69   Kuiper, H A et al (1999) 'Adequacy of Methods for Testing the Safety of Genetically Modified Foods', *The Lancet*, London, Vol 354, 16 October, pp1315–1316.

70   Fenton, B et al (1999) 'Differential Binding of the Insecticidal Lectin GNA to Human Blood Cells,' *The Lancet*, London, Vol 354, 16 October, pp1354–1355.

71   Kuiper et al (1999) op cit.

72   *The Lancet* (1999) 'GM Food Debate', London, Vol 354, 13 November, pp1725–1729.

73   Ewen, S and Pusztai, A (1999) Authors' Reply, *The Lancet*, London, Vol 354, 13 November, pp1726–1727.

74   Horton, R (1999) 'GM Food Debate', *The Lancet*, London, Vol 354, 13 November, p1729.

75   Brown, P (2002) 'British Scientists Turn on GM Food', *The Guardian*, London, 5 February, p7.

76   The Royal Society (2000) *Genetically Modified Plants for Food Use and Human Health – An Update*, London, February, p3.

77   Ibid; The Royal Society, *Safety Checks for GM Foods Must be Better, Says Royal Society*, Press Release, London, 4 February.

78   Gasson, M and Burke, D (2001). 'Scientific Perspectives on Regulating the Safety of Genetically Modified Foods', *Nature Reviews Genetics* 2, March, pp217–222.

79   Ibid.

80   Pusztai, A (2002) 'Can Science Give us the Tools for Recognising Possible Health Risks of GM Food'? *Nutrition and Health*, Vol 16, pp73–84.

81   Smith, J (2002) *Letter to Dr Pusztai*, London, 26 March.

82   Ewen, S (2002) *Re: The Royal Society GM Plants Report*, Email to Josephine Craig, 14 February.

83   Fares, N and El-Sayed, A (1998) 'Fine Structural Changes in the Ileum of Mice Fed on δ Endotoxin-Treated Potatoes and Transgenic Potatoes', *Natural Toxins*, Vol 6, pp219–233.

84   The Royal Society of Canada (2001) Expert Panel on the Future of Food Biotechnology, Ottawa, px; The Royal Society of Canada (2001) *Expert Panel Raises Serious Questions About the Regulation of GM Food*, Press Release, Ottawa, 5 February.

85   Ibid.

Chapter 7 Stars in their Eyes

1   Pribyl, L (1992) *Comments on Biotechnology Draft Document, 27 February*, Draft, 6 March.

2   Verakis, D (1998) Speaking on GMTV, 10 August, 6.25 am.

3   Barling, D and Henderson, R (2000) *Safety First? A Map of Public Sector Research into GM Food and Food Crops in the UK*, Centre for Food Policy, Thames Valley University, London, p50.

4   Diamand, E (2001) *The Great Food Gamble – An Assessment of Genetically Modified Food Safety*, Friends of the Earth, London, May; p5.

5   Science and Technology Committee (1999) *Scientific Advice to Government: Genetically Modified Food*, Memorandum Submitted by the Rowett Research Institute, The Stationery Office, London, 8 March, p39.

6   Food First (2002) *Biotech Industry Lobby Group Attacks Zambian Pro-Food-Rights Group*, Press Release, Oakland California, 15 October; Pusztai, A (2001) *Genetically Modified Foods: Are they a Risk to Human/Animal Health?*, June, http://actionbioscience.org/biotech/pusztai.html

7   The Pew Initiative on Food and Biotechnology (2002) *GM Food Safety: Are Government Regulations Adequate?*, Washington.

8   Quoted in Clark, A (2001) '"Luddites" Get Some Ammunition', *The Toronto Star*, 12 March.

9   Domingo, J (2000) 'Health Risks of GM Foods: Many Opinions but Few Data', *Science*, 9 June, Vol 288, No 5472, pp1748–1749.

10   Schapiro, M (2002) 'Sowing Disaster?', *The Nation*, 10 October.

11   National Research Council (2002) *Environmental Effects of Transgenic Plants: The Scope and Adequacy of Regulation*, Committee on Environmental Impacts Associated with the Commercialization of Transgenic Plants, Board on

Agriculture and Natural Resources, Division on Earth and Life Studies, National Academy Press, Washington, p49.

12   The Pew Initiative on Food and Biotechnology (2002) op cit.

13   Martineau, B (2000) *First Fruit – The Creation of the Flavr Savr™ Tomato and the Birth of Biotech Food*, McGraw Hill, New York, p7.

14   Martineau (2000) op cit, p9.

15   Ibid, p45.

16   Ibid, pp56–58.

17   Diamand, E (2001) *The Great Food Gamble*, Friends of the Earth, London, May, p17.

18   Martineau (2000) op cit, pp62, 90.

19   Ibid, p186.

20   Lumpkin, M (1992) Memorandum to Bruce Burlington, Re: *The Tomatoes That Will Eat Akron*, 17 December.

21   Sheldon, A (1993) *SUBJECT: Use of Kanamycin Resistance Markers in Tomatoes*, Memorandum to the CFSAN Biotechnology Co-ordinator, 30 March.

22   Martineau (2000) op cit, p99.

23   Ibid, p42.

24   Pribyl (1992) op cit.

25   Department of Health and Human Services, Food and Drug Administration (1992) 'Statement of Policy: Foods Derived from New Plant Varieties', *Federal Register*, Part IX, Vol 57, N 104, 29 May, pp22985, 22989–22990.

26   http://www.biointegrity.org

27   Freese, B (2001) *The StarLink Affair*, For submission to the FIFRA Scientific Advisory Panel considering Assessment of Additional Scientific Information Concerning StarLink™ Corn Meeting on July 17–19, Friends of the Earth, Washington, p36.

28   MacRae, J (1992) *Subject: FDA Food Biotechnology Policy*, Memorandum for C Boyden Gray, Executive Office of the President, 21 March.

29   Kahl, L (1992) *Memorandum about the Federal Register Document 'Statement of Policy: Foods From Genetically Modified Plants'*, to James Maryanski, 8 January.

30   Smith, M J (1992) *Comments on Draft Federal Register Notice on Food Biotechnology*, Memorandum to Jim Maryanski, 8 January.

31   Shibko, S (1992) *Subject: Revision of Toxicology Section of the 'Statement of Policy: Foods Derived from Genetically Modified Plants'*, Memorandum to James Maryanski, 31 January.

32   Pribyl (1992) op cit.

33   Johnson, C (1992) *Comments on Draft Statement of Policy, 12 December 1991*, 8 January.

34   Martineau (2000) op cit, p 70, 144; http://www.biointegrity.org

35   National Research Council (2000) *Genetically Modified Pest-Protected Plants – Science and Regulation*, Committee on Genetically Modified Pest-Protected Plants, Board on Agriculture and Natural Resources, National Academy Press, Washington, pp62–63.

36   Martineau (2000) op cit, pp169–170.

37   Ibid, p149.
38   Ibid, pp152, 175.
39   Hines, F (1993) *FLAVR SAVR Tomato*, Memorandum to Linda Kahl, Consumer Safety Office, Biotechnology Policy Branch, 16 June.
40   Scheuplein R J (1993) Response to Calgene Amended Petition, Memorandum, 27 October.
41   Food and Drug Administration (1994) *The Flavr Savr™ Arrives*, Press Release, Washington, 18 May.
42   Martineau (2000) op cit, ppx, 236–237.
43   Ibid.
44   Maryanski, J (1991) *Points to Consider for Safety Evaluation of Genetically Modified Foods*, Supplemental Information, Office of Compliance, Division of Food Chemistry and Technology, 1 November.
45   Martineau (2000) op cit, pp236–237.
46   Ibid, p224.
47   Pusztai, A (2002) 'Can Science Give us the Tools for Recognizing Possible Health Risks of GM Food?', *Nutrition and Health*, Vol 16, pp73–84.
48   Pusztai (2001) op cit.
49   Pusztai (2002) op cit, pp73–84.
50   Hansen, M (2002) *Interview with Author*, October.
51   Anderson, L (1999) *Genetic Engineering, Food and Our Environment – A Brief Guide*, Green Books, Totnes, pp27–28.
52   Taylor, M and Tick, J (2001) *The StarLink Case: Issues for the Future*, Resources for the Future, Washington, October, pp14–15.
53   National Research Council (2000) op cit, p66.
54   The common food allergies are peanuts, soybeans, milk, eggs, fish, crustacea, wheat, and tree nuts all causing problems that can range from a mild reaction to an anaphylactic shock.
55   Food and Agriculture Organization/World Health Organization (2001) *Evaluation of Allergenicity of Genetically Modified Foods, Report of a Joint FAO/WHO Expert Consultation on Allergenicity of Foods Derived from Biotechnology, 22–25 January*, Rome, pp1–6.
56   Freese (2001) op cit, p7; Bucchini, L and Goldman, L (2002) *Food Allergy – Implications For Genetically Modified Food*, Pew Initiative on Food and Biotechnology, Washington, June, p7.
57   Royal Society of Canada (2001) *Expert Panel on Biotechnology*, p206.
58   Bucchini and Goldman (2002) op cit, p6.
59   Rapporteur's Summary (2000) *GM Food Safety: Facts, Uncertainties and Assessment*. OECD Edinburgh Conference on the Scientific and Health Aspects of Genetically Modified Foods, 28 February–1 March.
60   FAO/WHO (2001) op cit, pp1–15.
61   Gendel, S (1998) 'The Use of Amino Acid Sequence Alignments to Assess Potential Allergenicity of Proteins Used in Genetic Engineering', *Advances in Food and Nutrition Research*, Vol 42, June, p60.
62   Taylor and Tick (2001) op cit, p14.
63   Smyth, S (2002) 'Liabilities and Economics of Transgenic Crops', *Nature Biotechnology*, Vol 20, No 6, June, pp537–538.
64   National Research Council (2000) op cit, p65.

65   Ibid, pp63, 72.

66   Friends of the Earth (2000) *Contaminant Found In Taco Bell Taco Shells,* Press Release, Washington, 18 September.

67   Ibid.

68   Figures taken from numerous different reports and press articles on StarLink.

69   *Reuters News Service* (2001) 'StarLink Tainted Corn from 1999, 2000 Crops', Washington, 20 March.

70   Freese (2001) op cit.

71   Munro, M (2002) 'Genetic Threats Blowin' in the Wind: Scientists Warn Modified Crops Are "Escaping and Going Rogues"', *National Post,* 7 June.

72   Taylor and Tick (2001) op cit, p1.

73   FIFRA Scientific Advisory Panel (2000) *A Set of Scientific Issues Being Considered by the Environmental Protection Agency Regarding: Assessment of Scientific Information Concerning StarLink™ Corn,* Report of Scientific Advisory Panel Meeting, November 28, 2000, EPA Washington, 1 December.

74   Hagelin, J (2000) *Statement for the FIFRA Scientific Advisory Panel Open Meeting on StarLink Corn,* Arlington, Virginia, 28 November.

75   FIFRA Scientific Advisory Panel (2000) op cit, p13.

76   www.cdc.gov/nceh/ehhe/Cry9cReport/executivesummary.htm

77   *CropChoice News* (2001) 'Environmental Group Points out Problems with StarLink Allergy Study', 18 June.

78   Muirhead, S (2000) 'Grain Elevators, Handlers Caught in Middle of StarLink Confusion', *Feedstuffs,* 6 November.

79   *Reuters* (2001) 'Canada: Banned GM StarLink Corn Mistakenly Fed to Animals, Reversal of Previous Government Statement', Ottawa, 16 March.

80   Hur, J (2001) 'Japan Groups Ask to Ban on US Biotech Corn Seed', *Reuters,* Tokyo, 20 April.

81   Friends of the Earth US (2000) *Foods Contaminated by Genetically Engineered US Corn Found in the United Kingdom,* Denmark, Press Release, Washington, 6 November.

82   Fuhrmans, V (2001) 'Aventis Fires Top Managers in Wake of StarLink Episode', *The Wall Street Journal,* 12 February, p12; Taylor and Tick (2001) op cit, p21.

83   Wolfson, R (2001) 'Biotech News', *Alive: the Canadian Journal of Health and Nutrition,* March.

84   FIFRA Scientific Advisory Panel (2001) *Assessment of Additional Scientific Information Concerning StarLink™ Corn,* in preparation for the Open Meeting, July 17–18, 2001, Virginia, EPA, 2 July, p5.

85   Finger, K (2001) *Testimony to the FIFRA Scientific Advisory Panel* (SAP) open Meeting, Virginia, 17 July.

86   FIFRA Scientific Advisory Panel (SAP) (2001) *Transcript of Open Meeting,* Assessment of Additional Scientific Information Concerning StarLink Corn, Virginia, 18 July.

87   Freese, B (2001) *Concerning the Revised Risks and Benefits Sections for Bacillus thuringiensis Plant-Pesticides,* Final Comments for Submission to the EPA, Friends of the Earth, Washington, 21 September.

88   Bernstein, L et al (1999) 'Immune Responses in Farm Workers after Exposure to *Bacillus Thuringiensis* Pesticides', *Environmental Health Perspectives*, Vol 107, No 7, July, pp575–582.

89   National Research Council (2000) op cit, pp112–113; quoting Hillbreck, A et al (1998) 'Effects of Transgenic *Bacillus thuringiensis* Corn-Fed Prey on Mortality and Development Time of Immature Chrysoperla Carnea (Neuroptera: Chrysopidea),' *Environmental Entomology*, Vol 27, pp480–487.

90   Vázquez-Padrón, R et al (1999) 'Intragastric and Intraperitoneal Administration of Cry1Ac Protoxin From *Bacillus thuringiensis* Induces Systemic and Mucosal Antibody Responses in Mice, *Life Sciences*, Vol 64, No 21, pp1897–1898.

91   Vázquez-Padrón, R et al (1999) 'Bacillus Thuringiensis Cry1Ac Protoxin is a Potent Systemic and Mucosal Adjuvant,' *Scandinavian Journal of Immunology* Vol 49, p578.

92   Vázquez-Padrón, R et al (2000) 'Cry1Ac Protoxin from *Bacillus thuringiensis* sp. *Kurstaki* HD73 Binds to Surface Proteins in the Mouse Small Intestine', *Biochemical and Biophysical Research Communications*, Vol 271, No 1, pp54, 58.

93   US Environmental Protection Agency (2001) *Bacillus thuringiensis Plant-Incorporated Protectants, Biopesticides Registration Action Document*, Washington, 29 September; US Environmental Protection Agency (2001) *Bacillus thuringiensis Plant-Incorporated Protectants, Biopesticides Registration Action Document*, Washington, 15 October.

94   Ibid.

95   http://www.mrc-lmb.cam.ac.uk/genomes/madanm/articles/dnashuff.htm

96   http://www.cf.ac.uk/biosi/staff/berry/chime/rtext.html

97   Kleter, G and Peijnenburg (2002) 'Screening of Transgenic Proteins Expressed in Transgenic Food Crops for the Presence of Short Amino Acid Sequences Identical to Potential, IgE – Binding Linear Epitopes of Allergens', *BMC Structural Biology*, 2:B.

98   Odvody, G et al (2000) *Aflatoxin and Insect Response of Near-Isogenic Bt and Non-Bt Commercial Corn Hybrids in South Texas*, Proceedings of the Aflatoxin/Fumonisin Workshop, Yosemite, 25–27 October, pp121–122.

99   Ibid; Odvody et al (date unknown) Funding Proposal to the USDA as part of the Initiative for Future Agriculture and Food Systems (IFAFS).

100   www.genewatch.org database

101   Block, T (2002) 'Pseudopregnancies Puzzle Swine Producer', *Iowa Farm Bureau Spokesman*, 29 April.

102   Kremer, R et al (2001) *Herbicide Impact on Fusarium spp. and Soybean Cyst Nematode in Glyphosate-Tolerant Soybean, Abstract*, University of Missouri, American Society of Agronomy Annual Meeting, Title Summary Number: S03-104-P: *Progressive Farmer* (2001) 'Glyphosate Weed Killer Benefits Soil Fungus', 3 January.

103   Losey, J et al (1999) 'Transgenic Pollen Harms Monarch Larvae', *Nature*, Vol 399, p214.

104   National Research Council (2000) op cit, p76; quoting Hansen, L and Obrycki, J (1999) *Non-target Effects of Bt Corn Pollen on the Monarch Butterfly* (Lepidoptera: Danaidae), Iowa, Iowa Sate University.

105   Dale, P et al (2002) 'Potential for the Environmental Impact of Transgenic Crops', *Nature Biotechnology*, Vol 20, June, p568.

106   National Research Council (2000) op cit, p37.

107   Dalton, R (2002) 'Superweed Study Falters as Seed Firms Deny Access to Transgene', *Nature*, Vol 419, 17 October, p655.

108   Dale et al (2002) op cit, p571.

109   ISAAA (2002) *Global GM Crop Area Continues to Grow and Exceeds 50 Million Hectares for First Time in 2001*, Press Release.

110   Caplan, R (2001) '"Bt" Means Big Trouble for Nation's Crops, Comment', *New York Times*, 30 August.

111   Science and Technology Committee (1999) *Scientific Advice to Government: Genetically Modified Food*, Minutes of Evidence, The Stationery Office, 8 March, p38.

112   Barling and Henderson (2000) op cit, pp60–66.

113   All the marketing consents given can be viewed on the www.genewatch.org database.

114   Diamand (2001) op cit, pP5, 34–37, 41.

115   Food Standards Agency (2001) *GM factsheet*, 15 October.

116   Advisory Committee on Novel Foods and Processes (ACNFP)(1999) *Annual Report 1998*, Ministry of Agriculture, Fisheries and Food and Department of Health, London, p1.

117   ACNFP (1996*) Government Gives Food Safety Clearance For Genetically Modified Tomatoes To Be Eaten Fresh*, Ministry of Agriculture, Fisheries and Food and Department of Health; London, 29 February.

118   Gilbert, R J (1993) *Advisory Committee on Novel Foods and Processes, Consultation Paper on the Use of Antibiotic Resistance Markers in Genetically Modified Organisms*, Public Health Laboratory Services, London, 30 November.

119   Ministry of Agriculture, Fisheries and Food (1995) *Three Novel Foods Cleared*, Press Release, London, 20 February.

120   ACNFP (1995) *ACNFP Annual Report 1994*, Ministry of Agriculture, Fisheries and Food and Department of Health, London, p40.

121   Bainbridge, J (2001) *The Use of Substantial Equivalence in the Risk Assessment of GM Food*, Submission to The Royal Society, 16 May.

122   National Research Council (2002) op cit, p 79.

123   The Medical Research Council (2000) *Report of a MRC Expert Group on Genetically Modified (GM) Foods*, London, June, p1.

124   Science and Technology Committee (1999) op cit, p39.

125   Mayer, S and Rutovitz, J (1996) *Trojan Tomatoes: Genetically Engineered for Delayed Softening or Ripening*, Final Draft – Unpublished, Greenpeace, London, 29 May, p6.

126   Padgette, S et al (1994) *Application to the United Kingdom Advisory Committee on Novel Foods and Processes for Review of the Safety of Glyphosate Tolerant Soybeans*, the Agricultural Group of Monsanto Company, Monsanto Europe, Belgium, 27 July, p10.

127   ACNFP (1999) *Annual Report 1998*, Ministry of Agriculture, Fisheries and Food and Department of Health, London, pp69–71.

128   Ciba Ceigy later merged with Sandoz to form Novartis. The agricultural divisions of Novartis and AstraZeneca later merged to form Syngenta.

129   ACRE (1995) *The Genetically Modified Organisms (Deliberate Release) Regulations 1992 – Application for Consent from Ciba Geigy to Market Genetically Modified Maize – C/F/94/11-03*, Annex Two, Advice Dated 22 May.

130   Marshall, S B (1995) *SNIF C For GM Maize, Ciba-Geigy Ltd: Notification No C/F/94/11-103*, Letter to Dr Parish, Ministry of Agriculture, Fisheries and Food, 15 May.

131   Marshall, S (1995) *SNIF C French Maize with Corn Borer Resistance and Glufosinate Herbicide Tolerance: Notification Number C/F/94/11-03*, Letter to Dr Parish, Department of the Environment, 19 May.

132   '*amp*' is an ampicillin-resistant gene; Anon (1995) *Genetically Modified Maize: Application for Consent to Market Under Directive 90/220/EEC Procedures*, Letter to Ranjini Rasaiah, MAFF, 20 June.

133   Anon (1996) Ciba-Geigy: *GM Maize: Antibiotic Resistance*, Letter to Ranjini Rasaiah, 29 January.

134   Pusztai, A (undated) Bt Maize Story, Notes.

135   European Commission (1996) *Report of the Scientific Committee for Animal Nutrition (SCAN) on the Safety for Animals of Certain Genetically Modified Maize Lines Notified by CIBA-GEIGY in Accordance with Directive 90/220/EEC for Feeding Stuff Use*, Directorate General VI, Agriculture, 13 December.

136   Rasaiah, R (1996) *Ciba Geigy Genetically Modified Insect Resistant Maize*, Fax to Professor Hammes, MAFF, 4 November.

137   Carr, S (2000) EU *Safety Regulation of Genetically-Modified Crops, Summary of a Ten-Country Study Funded by DGX11 Under its Biotechnology Programme*, Open University, Milton Keynes.

138   Kestin, S and Knowles, T (2000) *An Analysis of "the Chicken Study"*, Joint Proof of Evidence, Department of Clinical Veterinary Science, University of Bristol on behalf of Friends of the Earth, November, pp11–12.

139   Interviewed on Farming Today (2002) *Concern over Safety of GM Crops*, BBC, 27 April, 6.35 am.

140   http://www.defra.gov.uk/planth/pvs/gmrep1.htm

141   Aventis CropScience (2002) *T25 Maize: Feed Safety Assessment*, Submission to ACRE Open Hearing, London, February.

142   Friends of the Earth (2002) *Chardon LL Hearing – Under Regulation 16 of the Seed (National List of Varieties) Regulations 2001 Before Alun Alesbury*, London, 12 June, p1.

143   Millstone, E, Brunner, E and Mayer, S (1999) 'Beyond "Substantial Equivalence"', *Nature*, London, Vol 401, 7 October, pp525–526.

144   The Royal Society of Canada (2001) *Expert Panel on the Future of Food Biotechnology*, Ottawa, pix.

145   The Royal Society (2000) *Genetically Modified Plants for Food Use and Human Health – An Update*, London, February, p3; The Royal Society, *Safety Checks for GM Foods Must be Better, Says Royal Society*, Press Release, London, 4 February;

146   Bainbridge (2001) op cit.

147   Levidow, L (2002) 'Ignorance-Based Risk Assessment? Scientific Controversy Over GM Food Safety', *Science as Culture* Vol 11 No 1, March.

148   Mitchener, B (2002) 'EU Takes Steps to Require More Labelling for GMOs', *The Wall Street Journal*, 5 June.

149 Friends of the Earth (2002) *EC Commission Leaves GM Moratorium Decision to Member States*, Press Release, 17 October.
150 *Nature Biotechnology* (2002) 'Going With the Flow', Editorial, Vol 20, No 6, June, p527.
151 Smyth, S (2002) 'Liabilities and Economics of Transgenic Crops', *Nature Biotechnology*, Vol 20, No 6, June, pp537–538.
152 Coalition against BAYER-dangers (2002) *Bayer in GM Crop Contamination Scandals*, 30 August.
153 Iyer, V (2002) 'Crop Gene "Could Weaken Medicines"', *PA News*, 16 August.
154 University of Newcastle (2002) Technical Report on the Food Standards Agency Project G010008 'Evaluating the risks associated with using GMOs in human foods', Food Standards Agency, July.
155 Vidal, J (2002) 'GM Genes Found in Human Gut', *The Guardian*, London, 17 July.
156 Quoted in *The New York Times* (2001) '"Bt" Means Big Trouble for Nation's Crops', Comment, 30 August.

Chapter 8 Immoral Maize

1 *Nature Biotechnology* (2002) 'Going With the Flow', Vol 20, No 6, June, p527.
2 Platoni, K (2002) 'Kernels of Truth', *East Bay Express*, San Francisco, 29 May.
3 *BioDemocracy News* (2002) 'Frankencorn Fight: Cautionary Tales', No 37, January, p1.
4 National Research Council (2002) *Environmental Effects of Transgenic Plants: The Scope and Adequacy of Regulation*, Committee on Environmental Impacts Associated with the Commercialization of Transgenic Plants, Board on Agriculture and Natural Resources, Division on Earth and Life Studies, National Academy Press, Washington.
5 Quist, D and Chapela, I (2001) 'Transgenic DNA Introgressed into Traditional Maize Landraces in Oaxaco, Mexico', *Nature*, London, Vol 414, 29 November, p541.
6 University of California – Berkeley (2001) *Transgenic DNA Discovered in Native Mexican Corn, According to a New Study by UC Berkeley Researchers*, Press Release, 28 November; Quist and Chapela (2001) op cit, pp541–542.
7 Quist, and Chapela (2001) op cit, p542.
8 ETC Group (2002) *GM Fall-out from Mexico to Zambia: The Great Containment – The Year of Playing Dangerously*, Winnipeg, 25 October.
9 University of California – Berkeley (2001) op cit.
10 Chapela, I (2002) *Interview with Author*, 1 March.
11 BBC Radio 4 (2002) 'Seeds of Trouble', 7 January
12 Dalton, R (2001) 'Transgenic Corn Found Growing in Mexico', *Nature*, London, Vol 413, 27 September, p337.
13 Ferris, S (2002) 'Battle Lines Drawn in Mexico; Native Corn too Sacred to "Infect"?' *The Atlanta Journal and Constitution*, 28 February.

14   Quist and Chapela (2001) op cit, p541.
15   Dalton (2001) op cit.
16   Quist and Chapela (2001) op cit, p543.
17   Yang, S (2001) *Transgenic DNA Discovered in Native Mexican Corn, According to a New Study by UC Berkeley Researchers*, University of California Press Release, 29 November.
18   Press, E and Washburn, J (2000), 'The Kept University', *The Atlantic Monthly*, Vol 285, No 3, pp39–54.
19   See Rowell, A (1996) *Green Backlash – Global Subversion of the Environment Movement*, Routledge, London, New York.
20   Murphy, M (2001) Subject: *Mexican Corn – the new Starlink-Monarch-Mutant Scare Story*, Posted on the AgBioView site on the 29 November.
21   Smetacek, A (2001) Subject: *Ignatio Chapela – Activist FIRST, Scientist Second*, posted on the AgBioView site on 29 November.
22   MacGregor, B (2001) *Re: Genetically Modified Material Found in Mexican Corn*, posted on AgBioView list on 30 November.
23   Mann, C (2002) 'Has GM Corn "Invaded" Mexico?', *Science*, Vol 295, p1617, 1 March.
24   Kinderlerer, J (2001) *Regarding AGBIOVIEW: Chapela and Mexican corn, China, New Zealand support up, Lomborg, Peanut map*, Posted on the AgBioView list on 1 December.
25   http://www.foxbghsuit.com/wwwboard/messages/1168.html
26   See www.bivings.com
27   http://www.bivingsreport.com/search_view_full_article.php?article_id=73
28   Rowell, A (2002) 'Seeds of Dissent', *The Big Issue South West*, 15–21 April, pp16–17.
29   Ibid.
30   Copies of the emails provided to the author by the *The Ecologist*.
31   Rowell (2002) op cit.
32   Platoni (2002) op cit; Monbiot, G (2002) 'The Fake Persuaders', *The Guardian*, London, 14 May, p15.
33   Rowell (2002) op cit; Matthews, J (2002) 'Amaizing Disgrace', *The Ecologist*, London, Vol 32 No 4, May.
34   Bivings Group (2002) *Statement on the Ecologist story entitled 'Amaizing Disgrace'*, May.
35   Bivings, G (2002a) 'Bivings: We Condemn Online Vandalism', Letter to *The Guardian*, 12 June.
36   *BBC Newsnight* (2002) 'Row Over GM Crops – Mexican Scientist Tells Newsnight he was Threatened Because He Wanted to Tell the Truth', London, 7 June.
37   Bivings, G (2002b) 'The Maize Feud', *New Scientist*, 6 July.
38   Rowell (2002) op cit; Monbiot (2002) op cit; Matthews (2002) op cit; Bivings (2002a) op cit.
39   The original article was posted as the The Bivings Report (2002) *Viral Marketing: How to Infect the World*, 1 April.
40   Platoni (2002) op cit.
41   Bivings Report (2002) *Viral Marketing: The New Word of Mouth*, 1 November.

42   The Monsanto sites listed before the changes were Monsanto Corporate, Monsanto Africa; Monsanto France; Monsanto Fund; Monsanto India; Cornfacts; Monsanto Pakistan; Monsanto Spain; Monsanto UK; BioTech Basics; BioTech terms; Biotech Knowledge Centre; Monarch Info; Report on Sustainable Development and Teaching Science.

43   See cffar.org

44   Philipkoski, K (2002) 'A Dust-Up Over GMO Crops', *Wired News*, 12 June.

45   *The New York Times* (1999) 'Biotech Companies Take on Critics of Gene-Altered Food', 12 November; *The Wall Street Journal* (1999) 'Monsanto Fails Trying to Sell Europe on Bioengineered Food', 11 May.

46   Prakash, C (2002) *Interview with Author*, April.

47   Smetacek, A (2000) *A Plea to Stop Eco-Terror*, 21 July.

48   Rowell (2002) op cit.

49   Del Porto, D (2002) *Interview with Author*, Bivings, April.

50   http://www.v-fluence.com/about/team.html.

51   Bivings Group (2001) *A Look Into the Future of Online PR*, January.

52   Byrne, J (2001) *Protecting Your Assets: An Inside Look at the Perils and Power of the Internet*, a Presentation to the Ragan Communications Strategic Public Relations Conference, V-Influence, 11 December.

53   ETC Group (2002) *GM Pollution in the Bank? Time for "Plan B"*, News Release, Winnipeg, 4 February; Magallon Larson, H (2002) *Interview with Author*, 5 March.

54   Mann, C (2002) 'Has GM Corn Invaded Mexico?' *Science*, Vol 295, 1 March, p1617; Magallon Larson, H (2002) op cit.

55   *Transgenic Research* (2002) 'No Credible Evidence is Presented to Support Claims that Transgenic DNA was Introgressed into Traditional Maize Landraces in Oaxaca, Mexico', Editorial, 11, ppiii–v.

56   Hansen, M (2002) *Communication with Author*, June.

57   EPA (2001) *Bt Plant-Pesticides Biopesticides Registration Action Document*, Washington, 15 October, pIIA3.

58   Posted on the NGIN website on 27 February, 2002; http://www.ngin.org.uk

59   Rowell (2002) op cit.

60   Lepkowski W (2002) 'Biotech's OK Corral', *Science and Policy Perspectives*, No 13, 9 July.

61   Ibid.

62   ETC Group (2002) *UnNatural Rejection? The Academic Squabble Over Nature Magazine's Peer-Reviewed Article is Anything but Academic*, News Release, Winnipeg, 19 February.

63   Avery, A (2002) *Joint Statement from Scientists?* 21 February.

64   Prakash, C (2002) *Joint Statement of Scientific Discourse in Mexican GM Maize Scandal*, 24 February.

65   *Nature* (2002) 'Editorial Note', 4 April.

66   Webber, J (2002) *Interview with Author*, 4 April.

67   Campbell, P (2002) Letter to *The Guardian*, 15 May.

68   Clarke, M (2002) '*Suggestion on Mexican Maize Article*,' Email to Kate O'Connell, 10 June.

69   Platoni (2002) op cit.
70   Chapela, I (2002) *And Yet it Moves*, Letter to *The Guardian*, 24 May.
71   Suarez, A et al (2002) 'Correspondence', *Nature*, Vol 417, 27 June, p897.
72   *BBC NewsNight* (2002) 'Row Over GM Crops – Mexican Scientist Tells Newsnight he Was Threatened Because He Wanted to Tell the Truth', London, 7 June; Meek, J (2002) 'Science Journal Accused Over GM Article', *The Guardian*, London, 8 June.
73   Metz, M and Fütterer, J (2002) 'Suspect Evidence of Transgenic Contamination', *Nature*, 4 April; Kaplinsky, N et al (2002) 'Maize Transgene Results in Mexico Are Artefacts', *Nature*, 4 April.
74   Prakash, C (2002) *Joint Statement of Scientific Discourse in Mexican GM Maize Scandal*, 24 February.
75   http://www.public.iastate.edu/~iazelaya/Newsletter_Vol_1_No_3-Addendum.pdf
76   Apel, A (2001) *The Face of Terrorism*, Posting to AgBioView, 18 September.
77   Metz and Fütterer (2002) op cit.; Kaplinsky et al (2002) op cit.
78   Metz, M (2001) Correspondence, *Nature*, Vol 410, No 513, 29 March.
79   Chapela, I (2002) *Communication with the Author*, 15 April; quoting the ETH magazine "ETH-Life", 25 March
80   Chapela, I (2002) *Interview with Author*, 1 March.
81   Worthy, K, Strohman, R and Billings, P (2002) Correspondence, *Nature*, Vol 417, 27 June, p897.
82   Gee, H (2002) 'Food & the Future', *Nature*, Vol 418, 8 August, p667.
83   Metz, M and Fütterer, J (2002) Correspondence, *Nature*, Vol 417, 27 June, pp897–898; Kaplinsky, N (2002) Correspondence, *Nature*, Vol 417, 27 June, p898.
84   Quist, D and Chapela, I (2002) Brief Communications, *Nature*, 4 April.
85   AgBioWorld.org (2002) Mexican Maize Resource Library.
86   Brown, P (2002) 'Mexico's Vital Gene Reservoir Polluted by Modified Maize', *The Guardian*, London, 19 April.
87   Clover, C (2002) '"Worst Ever" GM Crop Invasion', *The Daily Telegraph*, 19 April.
88   Abate, T (2002) 'Hot Seat May Cool for Berkeley Prof: Mexican Scientists Reportedly Confirm his Findings of Engineered Corn in Maize', *The San Francisco Chronicle*, 26 August.
89   Rosset, P (2002) *Open Letter to Nature*, October; Food First (2002) *Nature Refuses to Publish Mexican Government Report Confirming Contamination of the Mexican Maize Genome by GMOs*, Press Release, Oakland, 24 October; ETC Group (2002) *GM Fall-out from Mexico to Zambia: The Great Containment The Year of Playing Dangerously*, Winnipeg, 25 October.
90   Prakash, C (2002) *Joint Statement of Scientific Discourse in Mexican GM Maize Scandal*, 24 February.
91   *Nature Biotechnology* (2002) op cit, p527.
92   Laidlaw, S (2001) 'Starlink Fallout Could Cost Billions', *The Toronto Star*, Toronto, 9 January.
93   Pearce, F (2002) 'The Great Mexican Maize Scandal', *New Scientist*, London, 15 June.

94   McGuire, D (2002) *Farmer Choice – Customer First When it Comes to GM Crops*, Presentation to 2002 Annual Convention of the American Corn Growers Association on 9 March; Washington, 13 March.

95   Villar, J (2001) *GMO Contamination – Around The World*, Friends of the Earth International, Amsterdam; Hager, N (2002) *Seeds of Distrust*, Craig Potton, Nelson, pp12–20.

96   Friends of the Earth (2002) *Manufacturing Drugs and Chemicals in Crops Biopharming Poses New Risks to Consumers, Farmers, Food Companies and the Environment*, Washington DC, July, Executive Summary; IPS (2002) 'Host for "Pharm Crop" Experiments,' Carmelo Ruiz-Marrero, Puerto Rico, 29 October.

97   Committee on Environmental Impacts Associated with Commercialization of Transgenic Plants of the National Academy of Sciences (2002) *Environmental Effects of Transgenic Plants: The Scope and Adequacy of Regulation*, National Academy Press, p68.

98   Schapiro, M (2002) 'Sowing Disaster?', *The Nation*, 10 October.

99   Friends of the Earth (2002) *Drugs And Chemicals Will Contaminate Food Supply Concludes New Report*, Press Release, Washington, 11 July.

Chapter 9 Science for Sale

1   Quoted by Eaglesham, J (2002) 'UK "Bad at Celebrating" Scientific Leadership', *The Financial Times*, London, 21 October.

2   Blair, T (2002) *Science Matters*, Speech to The Royal Society, London, 23 May.

3   Dickinson, A (2001) *Response to Questions from EFRA Select Committee*, 4 November.

4   Dickinson, A (1999) *Comments in BSE Epidemiology Session and Some Associated Documents*, 6 December.

5   Council for Science and Technology (2000) *Technology Matters – Report on the Exploitation of Science and Technology by UK Business*, London, February, Annex A, p26.

6   Barling, D and Henderson, R (2000) *Safety First? A Map of Public Sector Research into GM Food and Food Crops in the UK*, Centre for Food Policy, Thames Valley University, London, Discussion Paper 12, pp5, 6.

7   Halsey, A H (1995) *Decline of Donnish Dominion: The British Academic Professions in the Twentieth Century*, Oxford University Press, Oxford, p302; quoted in Howells, J et al (1998) *Industry-Academic Links in the UK*, Report for the Higher Education Funding Councils of England, Scotland and Wales, PREST, University of Manchester, December, p12.

8   Gristock, J and Senker, J (1999) *Public Science and Wealth Creation in Britain*, an Information Booklet Prepared for the British Council by SPRU Science and Technology Policy Research, University of Sussex, Brighton.

9   Ibid; Barling and Henderson (2000) op cit, pp14–15.

10   see http://www.foresight.gov.uk/

11   Taylor, I (1995) Quoted in Monbiot, G (200) *Captive State – The Corporate Takeover of Britain*, London, Macmillan, p285.

12   http://www.dti.gov.uk/ost/aboutost/index.htm

13   Quoted by Rutherford, J (2001) 'Uncreative Friendship', *The Times Higher Educational Supplement*, 27 April, p16.

14   Blair, T (1998), Foreword in *Our Competitive Future Building the Knowledge Driven Economy*, Department of Trade and Industry, London, December, p5.

15   Department of Trade and Industry (1998), *Our Competitive Future: Building the Knowledge Driven Economy*, London, December, p6.

16   Department of Trade and Industry (1998) *Science and Innovation at the Heart of the Knowledge Driven Economy – Mandelson*, Press Release, London, 16 December.

17   Howells et al (1998) op cit, p9.

18   Environmental Audit Select Committee (1999) *Genetically Modified Organisms and the Environment: Co-ordination of Government Policy*, Fifth Report, London, May.

19   Royal Society (1999) *The Royal Society Supporting Organisations*, 17 June.

20   Environmental Audit Select Committee (1999), op cit.

21   Barling and Henderson (2000) op cit, p19.

22   Gillard, M S and Flynn, L (1999) 'Key GM Gene is Owned by Monsanto', *The Guardian*, 17 February, p1.

23   Dillon, J (2002) 'Sainsbury is Attacked for GM Share "profits"', *The Independent on Sunday*, 26 May.

24   Lord Sainsbury (1999) *Biotechnology Clusters*, Report Prepared by a Team Led by the Minister for Science, Department of Trade and Industry, London, August, p2.

25   Department of Trade and Industry (2000) *Funding Boost For Biotechnology Enterprise*, Press Release, London, 24 July.

26   For more information see http://www.foresight.gov.uk/

27   The six were Communication in the Food Chain; Spreading Best Practice; Unlocking the Potential of Industrial Crops; Food's Contribution to Health in the Future; the Debate on the Use of Technology in the Food Chain and the Future Skill Needs of the UK's Food Chain and Crops for Industry Sectors.

28   www.foresight.gov.uk/

29   Food Chain & Crops for Industry Foresight Panel (2000) *Debate on The Use of Technology in the Food Chain*, 23 February, Department of Trade and Industry, London.

30   Council for Science and Technology (2000) *Technology*, London, February, pp8, 12.

31   Department of Trade and Industry (2000) *Excellence and Opportunity – a Science and Innovation Policy for the 21st Century*, London, July, p8, 28–29.

32   10 Downing Street (2001) *120 Million for Science Knowledge Transfer Funding*, Press Release, 2 October.

33   Department of Trade and Industry and Office for Science and Technology (2001) *Forward Look 2001*, 17 December.

34   Brown, G, Hewitt, P and Morris, E (2002) Preface to *Investing in Innovation – a Strategy for Science, Engineering and Technology*, HM Treasury, DTI, and Department for Education and Skills, The Stationery Office, London, p1.
35   Dickinson, A (2001) *Letter to the Chairman of the Agriculture Select Committee*, House of Commons, 25 January.
36   Gristock and Senker (1999) op cit.
37   http://www.bbsrc.ac.uk/about/gov/structure.html#council
38   http://www.bbsrc.ac.uk/about/gov/members.html#council
39   Office of Science and Technology (2000) *Biotechnology in the UK – A Scenario for Success in 2005*, A Report Commissioned by Dr John Taylor, Director-General of the Research Councils, 17 November.
40   Ibid.
41   Ibid.
42   BBSRC (2000) *BBBSC Welcomes White Paper Recognition for Basic Science and Exploitation*, Press Release, Swindon, 28 July.
43   BBSRC (2001) *Annual Report 2000–2001*, Swindon, 11 July, p23.
44   BBSRC (1999) *Strategic Plan 1999–2004*, Swindon, p1.
45   For more details see http://www.rri.sari.ac.uk/
46   Scottish Crop Research Institute (1999) *Annual Report 1998/99*, Dundee, p43; www.scri.sari.ac.uk; www.mrsltd.com.
47   BBSRC (1998) *BBSRC Welcomes Industrial Alliances at John Innes Centre*, Press Release, Swindon, 16 September.
48   Lamb, C and Baulcombe, D (2000), in *John Innes Centre and Sainsbury Laboratory, Annual Report 1999–2000*, Directors Report, Norwich, p3.
49   John Innes Centre (2001) *Laying the Foundation for More Science at the John Innes Centre*, Press Release, Norwich, 16 July.
50   Allen-Stevens, T (2002) 'Syngenta Quits Genome Centre', *Farmers' Weekly Interactive*, 19 September.
51   Meek, J (2002) 'Cloned Pigs Give Vital Boost to Future of Transplants,' *The Guardian*, London, 3 January, p1; BBC (2002) News at Ten, London, 3 January; PPL Therapeutics (2002) *World's First Announcement of Cloned "Knock-Out" Pigs*, Press Statement, Edinburgh, January; Naik, G (2002) 'Quick Publishing of Research Breakthroughs Can Lift Stock, but May Undermine Science', *The Wall Street Journal*, 28 January.
52   Ibid.
53   BBC World Service (2000) 'Science Fact or Fraud?', London, 15 September.
54   Irwin, A (1999) 'GM Food Scares "Risk Britain's Lead in Science"', *The Daily Telegraph*, London, 19 May, pp1, 7.
55   Nichols, M (1999) 'Pusztai Accused of Costing Country Millions over GM', *Press and Journal*, Aberdeen, 26 February.
56   Horton, R (1999) Quoted in draft of article for *The Guardian* that finally was published on November 1, 26 October.
57   The Royal Society of Canada (2001) *Expert Panel on the Future of Food Biotechnology*, Ottawa, p216.
58   The Royal Society of Canada (2001) *Expert Panel Raises Serious Questions About the Regulation of GM Food*, Press Release, Ottawa, 5 February.

59   Royal Society of Canada (2001) *Expert Panel on Biotechnology*, p217.
60   Wynne, B (1999) Quoted in draft of article for *The Guardian*, 26 October.
61   IPMS (1999) *What Future R&D?*, London, November.
62   IPMS (2001) Scientists Speak For Themselves, *What Future R&D*, March.
63   Rowell, A (2001), Notes on Corruption of Scientific Integrity? – The Commercialization of Academic Science conference, May; Meeting Report at http://www.cafas.org.uk/meetingreport.htm.
64   Thompson, J, Baird, P and Downie, J (2001) *The Olivieri Report – The Complete Text of the Report of the Independent Inquiry Commissioned by the Canadian Association of University Teachers*, James Lorimer and Company, Toronto; http://www.caut.ca/english/issues/acadfreedom/Olivieri%20 Inquiry%20Report.pdf
65   Rowell (2001) op cit.
66   Thompson, Baird and Downie (2001) op cit, p17.
67   Fine, P, (2001) 'Toronto Pulls Job Offer from Prozac Critic', *The Times Higher Education Supplement*, London, 27 April, p1.
68   Boseley, S (2002) 'It Said the Drug Was the Best Thing Since Sliced Bread. I Don't Think it is', *The Guardian*, London, 7 February, p4.
69   Boseley, S (2002) ' Scandal of Scientists Who Take Money for Papers Ghostwritten by Drug Companies', *The Guardian*, London, 7 February, p4.
70   See Shenk, D (1999) Money + Science = Ethics Problems on Campus, *The Nation*, 22 March.
71   http://www.doctorsintegrity.com/exist/kern.htm
72   *The Lancet* (2001) 'The Tightening Grip of Big Pharma', Editorial, London, Vol 357, No 9263, 14 April, p1141.
73   Dalton, R (2002) 'Superweed Study Falters as Seed Firms Deny Access to Transgene', *Nature*, Vol 419, 17 October, p655; Brown, P (2002) 'Scientists Shocked at GM Gene Transfer', *The Guardian*, London, 15 August.
74   Ramsey, S (2001) 'Controversial MMR-Autism Investigator Resigns from Research Post', *The Lancet*, Vol 358, 8 December, p1972.
75   The Policy Commission on the Future of Farming and Food (2002) *Farming & Food – A Sustainable Future*, January, p90, 100–101.
76   Boseley (2002) op cit, p4
77   Quoted in Shenk (1999) op cit.
78   Lightfoot, L (2000), 'Scientists "Asked to Fix Results for Backer"', *The Daily Telegraph*, London, 14 February.
79   Press, E and Washburn, J (2000), 'The Kept University', *The Atlantic Monthly*, (Part Two) Vol 285, No 3, pp39–54

Chapter 10 Whitehall Whitewash

1   Food Standards Agency (2002) *The First Two Years*, London, quoting Leather on the inside front cover.
2   Consumers' Association (2002) *Consumers Call For Full Labelling of GM Foods*, Press Release, London, 4 June.

3   House of Commons (2002) *Draft Ministry of Agriculture, Fisheries and Food (Dissolution) Order 2002*, Fifth Standing Committee on Delegated Legislation, London, 22 January.

4   MAFF (2001) *Department for Environment, Food and Rural Affairs*, London, Press Release, 9 June.

5   Ibid.

6   Lawrence, F (2001) 'This is Our Chance for Safer Food', *The Guardian*, London, 27 June.

7   Brown, D (2001) 'Mixed Rural Welcome for Beckett Super-Ministry', *The Daily Telegraph*, London, 11 June.

8   http://www.cullmaff.com/

9   http://www.beyless.freeserve.co.uk/ArofWe11.htm

10   http://www.archive.official-documents.co.uk/document/maffdh/5113/ax8.htm

11   Bender, B (2001) *Evidence to the Select Committee on Environment, Food and Rural Affairs*, London, 14 November.

12   Ibid.

13   *The Daily Telegraph* (2002) 'Who is Blamed for the Foot and Mouth Crisis?' London, 23 July.

14   Booker, C and North, R (2001) 'Not the Foot and Mouth Report', *Private Eye*, London, November, p23.

15   Bender (2001) op cit.

16   Kemp, Sir P (2001) *Defra and the Structure of Whitehall*, 12 October.

17   Institute of Public Policy Research (IPPR) (2000) *Remodelling Whitehall – Project Proposal*, London, December.

18   Kirby, A (2001) 'New Green Ministry Faces Tests', *BBC News Online*, 11 June.

19   Robertson, J (2001) 'Moving Deckchairs or Changing Course – Will Defra Make A Difference?', *Ecos Magazine*, Volume 22, Number 2, September.

20   Vidal, J (2002) 'Fresh Fields', *The Guardian*, Society Section, London, 12 June.

21   Gibson, A (2002) *The Future of Farming in Devon*, Talk at the Barn Theatre, Dartington, Devon, 26 October.

22   House of Commons (2002) op cit.

23   Defra (2002) *Departmental Report 2002*, Presented to Parliament by the Secretary of State for Environment, Food and Rural Affairs, London, May.

24   The Environment, Food and Rural Affairs Committee (2002) *The Departmental Annual Report 2002*, Sixth Report, House of Commons, The Stationery Office, London, pp8–12.

25   Uhlig, R (2002) 'Defra Report "Riddled with Inaccuracies and Poetry,"' *The Daily Telegraph*, London, 18 July.

26   Porritt, J (2002) 'Can She Sustain it?', *The Guardian*, London, 1 October, p19.

27   Robertson (2001) op cit.

28   Food Standards Agency (2002) *The First Two Years*, London, p1.

29   Beckett, M (2001) *Evidence to the Select Committee on Environment, Food and Rural Affairs*, London, 14 November.

30   http://www.nfbg.org.uk/

31   Rowell, A (2001) 'Oganicised Crime', *The Ecologist*, Vol 31, No 1, February, p33.

32   *Private Eye* (2000) 'Man in the Eye', No 994, 28 January.

33   Rowell (2001) op cit.

34   Soil Association (2000) *Soil Association Responds to the Food Standards Agency*, Press Release, Bristol, 2 September.

35   http://www.sirc.org/news/guidelines.shtml

36   See www.sirc.org; or MCM research at www.i-way.co.uk/~mcm/index.html.

37   Ferriman, A (1999) 'An End to Health Scares?' *BMJ*, Vol 319, No 716, 11 September.

38   http://www.sciencemediacentre.org/aboutus/scienceadvisorypanel.html

39   Pearson, H (2001) 'Science Media Centre Launches – London Hub Hopes to Nurture Science-Media Affairs', *Nature*, 8 November.

40   McKie, R (2002) The GM Conspiracy – Lobby group 'Led GM thriller Critics,' *The Observer*, London, 2 June.

41   Norfolk Genetic Engineering Network (NGIN) (2002) *Science Media Centre Exposed – NGIN*, 2 June, http://ngin.icsenglish.com

42   Krebs, Sir J (2000) Interview on *Country File*, BBC TV, 3 August.

43   *The Guardian* (2000) 'Krebs Views "Extreme" Says Food Safety Authority Head', London, 14 September.

44   Edwards, R (2002) 'Organic Food Might Reduce Heart Attacks', *New Scientist*, London, 14 March.

45   Lean, G (2002) 'Organic Farming Shunned by Food Watchdog', *The Independent on Sunday*, 3 November.

46   Lang, Professor T (2000) Interviewed on the *Food Programme*, BBC Radio 4, 17 September.

47   Rowell (2001) op cit, p34.

48   Charman, K (1999) Saving the Planet with Pestilent Statistics, *PR Watch*, Fourth Quarter; prwatch.org

49   See Avery's article called 'Warning: Organic and Natural Foods may be Hazardous to Your Health', 2000, 2 May at www.guestchoice.com

50   Cummins, R and Lilliston, B (1999) 'Organics Under Fire: The US Debate Continues', *Food Bytes*, 25 January, No 16.

51   Rampton, S and Stauber, J (2000) 'The Usual Suspects: Industry Hacks Turn Fear on Its Head', *PR Watch*, Third Quarter, pp4, at www.prwatch.org

52   Rowell, A. (1996) *Green Backlash – Global Subversion of the Environmental Movement*, Routledge, London, p328. For example, in 1994, the Institute for Economic Affairs published a report 'Global Warming: Apocalypse or Hot Air?' by Roger Bate and Julian Morris. In the chapter on climate change science around half of the references cited were from known climate sceptics, many who received funding from the fossil fuel lobby. It was dismissed by the experts as 'uninformed and hard to take seriously'.

53   ESEF (1996) *The Global Warming Debate*, European Science and Environment Forum, London

54   Stauber, J and Rampton, S (2001) *Trust Us, We're Experts: How Industry Manipulates Science and Gambles with Your Future*, Tarcher/Putnam, pp242–243, January; Burson-Marsteller (undated) *Scientists for Sound Public Policy – Assessment Project and Symposium*, Bates Numbers 2028363773–2028363791; Ong, E and Glantz, S (2000) 'Tobacco Industry Efforts Subverting International Agency for Research on Cancer's Second-Hand Smoke Study', *The Lancet*, Vol 355, pp1253–1258, 8 April; Ong, E and Glantz, S (2001) 'Constructing "Sound Science" and "Good Epidemiology": Tobacco, Lawyers, and Public Relations Firms', *American Journal of Public Health*, Vol 19, No 11, pp1749–1757.

55   IEA (1999) *Londoners Demand Regulation of Potentially Deadly Organic Food*, Press Release, London, 16 August.

56   Morris, J (2002) 'What is Sustainable Development and How Can We Achieve it'? *In Sustainable Development: Promoting Progress or Perpetuating Poverty?* Profile Books, London.

57   Morris, J and Bate, R (1999) *Fearing Food: Risk, Health & Environment*, Butterworth Heinemann, Oxford.

58   See www.bioindustry.co.uk

59   Trewavas, A (1999) 'Much Food, Many Problems', *Nature*, 18 November.

60   GeneWatch UK (2002) *GeneWatch UK Welcomes MEPs' Vote to give Consumers Choice About GM Food*, Press Release, 3 July; GeneWatch UK (2002) *House of Lords European Union Committee Evidence*, February.

61   http://www.foodstandards.gov.uk/science/sciencetopics/gmfoods/gm_labelling

62   National Economic Research Associates (2001) *Economic Appraisal of Options for Extension of Legislation on GM Labelling*, A Final Report for the Food Standards Agency, London, May.

63   GeneWatch UK (2002) op cit.

64   GeneWatch UK (2002) *Proposed EU Regulations on the Traceability and Labelling of GM Food and Animal Feed,* European Parliamentary Briefing Paper, Number One, Tideswell, February, p1.

65   GeneWatch UK (2002) *The Labelling of GMOs and Their Derivatives in Food and Animals Feed*, European Parliamentary Briefing Paper, Number Two, Tideswell, June, pp2–3.

66   GeneWatch UK (2002) *GeneWatch UK Welcomes MEPs' Vote to give Consumers Choice About GM Food*, Press Release, 3 July.

67   Consumers Association (2002) op cit.

Chapter 11 Towards Safe Food and Public Interest Science

1   Berry, W (1995) *Another Turn of the Crank*, Counterpoint Press, Boulder, Colorado.

2   Gibson, A (2002) *The Future of Farming in Devon*, Dartington, Devon, 26 October.

3   Ibid.
4   Friends of the Earth (2001) *Response to the Policy Commission on the Future of Farming and Food*, London, October, p3.
5   www.riverford.co.uk
6   Watson, G (2002) *The Future of Farming in Devon*, Talk at the Barn Theatre, Dartington, Devon, 26 October.
7   Policy Commission on the Future of Farming and Food (2002) *Farming & Food – A Sustainable Future*, January, p16.
8   Food Ethics Council (2001) *After FMD: Aiming for a Values Driven Agriculture*, Southwell, p11.
9   CB Hillier Parker (1998) *The Impact of Large Foodstores on Market Towns and District Centres*, The Department for Environment, Transport and the Regions, Stationery Office, London, September.
10   Porter, S and Raistrick, P (1998) *The Impact of Out-Of-Centre Food Superstores on Local Retail Employment*, The National Retail Planning Forum, January.
11   Breed, C (1998) *Checking Out the Supermarkets: Competition in Retailing*, 1998, May.
12   Raven, H, Lang, T and Dumonteil, C (1995) 'Off Our Trolley?' *Food Retailing and the Hypermarket Economy*, Institute for Public Policy Research, London.
13   Sustain & Elm Farm Research Centre (2001) *Eating Oil – Food in a Changing Climate*, December, London.
14   Competition Commission (2000) quoted by Food Ethics Council (2001) op cit, p12.
15   Sustain & Elm Farm Research Centre (2001) op cit.
16   Ibid.
17   Blythmann, J (2002) 'Strange Fruit', *The Guardian*, Weekend Section, 7 September, pp20–24.
18   Sustain (undated) *Apples And Pears*, Factsheet.
19   Sustain (undated) *Salad Days*, Factsheet.
20   Blythmann (2002) op cit.
21   Lawrence, F (2002) 'Farmers Lose Out on Retail Sales', *The Guardian*, London, 9 September, p9.
22   *The Guardian* (2001) 'From Farm to Plate – A Sick Industry,' London, 28 February, p5.
23   Gibson (2002) op cit.
24   Policy Commission on the Future of Farming and Food (2002) op cit, pp17, 20, 23, 24, 77, 109.
25   Gibson (2002) op cit.
26   Sustain, UK Food Group, ActionAid (2002) *The CAP Doesn't Fit – Sustain and UK Food Group Recommendations for Reform of the Common Agricultural Policy*, London, July, p7.
27   Gibson (2002) op cit.
28   Hines, C and Shiva, V (2002) *A Better Agriculture is Possible: Local Food, Global Solution*, A Discussion Paper Prepared by the International Forum on Globalization & Research Foundation for Science, Technology and Ecology, June.

29   Wood, P and Hart, M (2002) *A Better CAP*, Family Farmers' Association, June, p2.
30   Food Ethics Council (2001) op cit, pp3–4.
31   Hines and Shiva (2002) op cit.
32   Vidal, J (2001) 'Global Trade Forces Exodus from Land', *The Guardian*, London, 28 February, p4; Friends of the Earth (2001) op cit, p5.
33   see http://www.farm.org.uk
34   Hetherington, P (2002) 'Hedges, Reed Beds, Meadows – and Meat Direct', *The Guardian*, London, 30 January, p7.
35   *Western Morning News* (2001) 'Foot and Mouth – How the West Country Lived Through the Nightmare', Plymouth, p17.
36   Food Ethics Council (2001) op cit, p12.
37   Coleman, J (2002) 'Future Lies in Better Branding', *Western Morning News*, Plymouth, Farming Section, pp4–5.
38   Oliver, J (2002) 'Local Food for Local People?' *Farmers Weekly*, Sutton, Vol 137, No 14, 4 October, p14.
39   Quoted in *Western Morning News* (2001) 'Level the Field on the Price of Food', Plymouth, 1 May, p15.
40   Trewin, C (2001) 'Rural Affairs Must be Top of the Agenda', *Western Morning News*, Plymouth, Westcountry Farming Section, 23 May, pp4–5; *Western Morning News* (2001) 'Praise for WMN's Buy Local Campaign', 17 December, p3.
41   Elliott, V (2002) 'Charles Calls on Shoppers to Rethink', *The Times*, London, 2 July, p2.
42   Gibson (2002) op cit.
43   *The Guardian* (2001) 'From Farm to Plate – A Sick Industry,' London, 28 February, p5.
44   Spedding, C (2000) *Animal Welfare*, Earthscan, London, p5; quoted by Food Ethics Council (2001) op cit, p20.
45   Soil Association (2002) *Annual Review 2002*, Bristol, p7.
46   Webster J (1995) *Animal Welfare: A Cool Eye Towards Eden*, Blackwell, Oxford, p170; quoted by Food Ethics Council (2001) op cit, p20.
47   Ross, D (2002) 'Glorious Foodie', *The Independent*, London, Review, pp4–5.
48   Rowell, A (2002) 'Happy as a Pig in Muck', *The Big Issue*, London, 18–24 February, pp14–15.
49   Bishop, S (2002) 'Wealden Farmers Network', *Oates Newsletter*, Common Cause Co-operative, Lewes, Vol 2, No 2, July.
50   Ibid.
51   Hetherington, P (2002) 'Burning Issues', *The Guardian*, London, 21 February, p17.
52   http://www.redtractortruth.com/
53   Nicolson-Lord, D (2001) 'The Future of Food: Safety First For Agriculture', *The Independent*, London, 7 July.
54   Soil Association (2002) *Annual Review 2002*, Bristol, p10.
55   Soil Association (2002) *Soil Association Welcomes Groundbreaking Report*, Press Release, Bristol, 29 January.

56 Organic Targets Campaign & Friends of the Earth (2002) *Government Organic Action Plan Welcomed – But Campaigners Call for Action Timetable*, Press Release, London, 29 July.

57 Sustain (2002) *Supermarkets Failing to Buy British Organic Produce*, Press Release, London, 26 July.

58 Lawrence, F (2002) 'Organic Sales Boom But Most Imported', *The Guardian*, London, 15 October, p8.

59 Ross, A (2002) *Organic Food Prices 2002 – Comparisons of Prices: Supermarkets; Farm Shops; Box Schemes; Farmers Markets*, University of West of England, pp1, 6.

60 Soil Association (2002) *New Initiative to Boost UK Organic Food*, Press Release, Bristol, 11 October.

61 Sustain (2001) *Obesity: Junk Food Advertising Ban To Protect Kids*, Press Release, London, 21st January.

62 see http://www.sustainweb.org

63 Lang, T and Rayner, G (eds) (2002) *Why Health is the Key to the Future of Food and Farming – A Report on the Future of Farming and Food*, Endorsed by the following organizations: Chartered Institute of Environmental Health; Faculty of Public Health Medicine of the Royal Colleges of Physician; National Heart Forum; UK Public Health Association; Supported by the Health Development Agency, p3.

64 Lang and Rayner (2002) op cit., p4.

65 Stirling, Dr A (2002) *Communication with Author*, October.

66 Wynne, B (1992) 'Uncertainty and Environmental Learning: Reconceiving Science and Policy in the Preventive Paradigm', *Global Environmental Change*, Vol 6, pp111–127; Wynne, B (2001) *Managing and Communicating Scientific Uncertainty in Public Policy*, Harvard University Conference on Biotechnology and Global Governance: Crisis and Opportunity, Kennedy School of Government, April; quoted in European Environment Agency (2001) *Late Lessons from Early Warnings: the Precautionary Principle 1896–2000*, Copenhagen, p185.

67 Policy Commission on the Future of Farming and Food (2002) op cit, p28, 101.

68 Stirling, A (1999) 'On Science, and Precaution in the Management of Technological Risk', Final Summary Report, *Technological Risk and the Management of Uncertainty Project*, European Scientific Technology Observatory, EC Forward Studies Unit, Brussels; quoted in European Environment Agency (2001) op cit, p185.

69 Royal Society of Canada (2001) *Expert Panel on Biotechnology*, pp198, 205.

Index